CHEMISTRY RESEARCH AND APPLICATIONS

WET ELECTROCHEMICAL DETECTION OF ORGANIC IMPURITIES

CHEMISTRY RESEARCH AND APPLICATIONS

Additional books in this series can be found on Nova's website at:

https://www.novapublishers.com/catalog/index.php?cPath=23_29&seriesp=
Chemistry+Research+and+Applications

Additional E-books in this series can be found on Nova's website at:

https://www.novapublishers.com/catalog/index.php?cPath=23_29&seriespe=
Chemistry+Research+and+Applications

CHEMISTRY RESEARCH AND APPLICATIONS

WET ELECTROCHEMICAL DETECTION OF ORGANIC IMPURITIES

F. MANEA

Nova Science Publishers, Inc.
New York

For permission to use material from this book please contact us:
Telephone 631-231-7269; Fax 631-231-8175
Web Site: http://www.novapublishers.com

Additional color graphics may be available in the e-book version of this book.

LIBRARY OF CONGRESS CATALOGING-IN-PUBLICATION DATA

Available upon request.
ISBN: 978-1-61668-661-1

Published by Nova Science Publishers, Inc. † New York

CONTENTS

PREFACE

Increased concentrations of organic pollutants in waters can result from the growth of certain industrial activities. In order to develop a fast electrochemical method to quantitatively determine the concentrations of such pollutants, new electrode materials are required. This review will focus on the preparation, characterization, and electrochemical behaviour of advanced carbon-based composite electrodes, i.e., expanded exfoliated graphite-epoxy, carbon nanofiber-expanded graphite-epoxy, expanded graphite-epoxy, and expanded graphite-polystyrene. Scanning Electron Microscopy (SEM) has been used to determine the physical distribution of the carbon particles within the insulating matrices epoxy or polystyrene. Expanded graphite-epoxy composites were used as the host material for a Cu or Ag-modified zeolite. Selected organic molecules were chosen to represent target pollutants in aqueous solutions, i.e., thiourea for sulfur-based impurities, glucose for carbohydrates, and urea for nitrogen-based impurities. In addition, phenol derivates as priority pollutant set by EU and by USEPA and oxalic acid as a result of the partial degradation of phenol derivates were selected as well. The electrochemical characterization and application of these composite electrodes were carried out in comparison with another alternative promising C-based electrode, commerical boron-doped diamond electrode. Cyclic voltammetry, linear-scan voltammetry, chronoamperometry, differential-pulsed voltammetry, and multiple-pulsed amperometry were used to determine the electroanalytical performances of the electrode materials. The zeolite-modified electrodes exhibit improved sensitivity for the detection of 4-chlorophenol and improved catalytic activity for the detection of urea and glucose. The presented advanced electrode materials and the selected electrochemical methods reveal that the fast determination of the concentrations has been achieved.

Chapter 1

INTRODUCTION

In general, the industrial activities led to increasing of pollutants content in wastewater, which further influences the water quality by the water cycle in the environment. Integrated water management usually requires monitoring of key pollutants that serve as indicators of acceptability for a specific use. An important pollutants class for wastewater monitoring is organics, usually expressed as a conventional Chemical Oxygen Demand (COD) parameter. Due to the chemical methods for evaluating COD are the time-consuming processes and require operating skills, a new alternative unconventional method for COD or direct organics content evaluation could be electrochemical and photoelectrochemical methods [1-7], especially direct electrochemical detection. An electrochemical assessement of organic load can be achieved as AOL (amperometric oxidability level) index [5]. AOL index can be expressed as the fast amperometric signal obtained in a certain potential range prior settled based on the voltammogram [5].

The electrochemical detection method of organics from water requires appropriate types of the electrode materials with suitable properties for their use as electrochemical detectors. Several characteristics are imposed, involving mainly high overvoltage for oxygen evolution in order to direct detection, the low and stable background current, and as possibly the cheapness of the working electrodes. However, the electrochemical assessment of COD parameter was reported under the oxygen evolution potential with BDD electrode at the potential value of +2.5 V/SCE [2]. Also, different types of electrodes were applied for the electrochemical assessment of COD parameter, i.e., undoped and F/Co-doped PbO_2, copper oxides based electrodes [3, 5-7]. A very fast method with good accuracy for COD assessment was achieved by photoelectrocatalytic technique using TiO_2 nanotube array electrode [4].

The organic load that is mainly responsible for COD parameter from wastewaters, imperiously includes various organics from: S - containing organics class (e.g.: thiourea); carbohydrates class (e.g.: glucose); N-containing organics class (e.g.: urea), which are found in biological waste. Also, the organics which are included into the lists of "priority pollutants", such as the phenol derivates class necessitates the quantitative evaluation. The oxidation potential of phenol derivates is function of substitution degree, group type and position. Because of the refractory property of phenol derivates, sometimes no complete mineralization occurred during the degradation process and intermediates or almost quite final products of carboxylic acids (e.g., oxalic acid) are formed. In this context, simultaneous or individual determination of oxalic acid in the presence of phenol derivates is desired. In addition, a growing interest is represented by the assessment of the simultaneous and selective determination of pollutants from different organics class responsible for COD, e.g., carbohydrates and phenol derivates.

Usual, a real water/wastewater contains various types of organics, which are responsible for organic load. For the electrochemical detection of COD parameter given by various types of organics, different situations can occur, i.e., the organics can interfere each other and the amperometic signal is diminished, or to have an cumulative effect at an oxidation potential at which to assure the electrochemical conditions to oxidize the organics with the low oxidation potential. To establish some aspects very important for the electroanalysis of organics at carbon-based electrode, number a studies have been carried out to investigate the electrochemical behaviour of different kinds of organic pollutants in aqueous phase.

Three types of advanced carbon-based electrode materials, i.e., boron-doped diamond, expanded graphite-epoxy/polystyrene and zeolite modified-expanded graphite-epoxy composite electrodes are presented to detect various types of organics in aqueous solution.

Chapter 2

BORON-DOPED DIAMOND ELECTRODE

Solid electrodes based on carbon receive growing interest as electrode material, because they exhibit many advantages, including good electrical conductivity, chemical inertness, wide potential range and low cost [1, 8]. In general, electron transfer rates at carbon surfaces are often slow, being affected by the origin and history of the carbon surface. All common carbon electode materials have the basic structure of a six-membered aromatic ring and sp^2 bonding, they differ in the relative density of the edge and basal planes at their surfaces [8]. Despite of their advantages, traditional carbon electrodes suffer some drawbacks, the most important being potential window for water stability and electrode fouling during the electrooxidation process. For the electroanalysis, the electrode fouling impose the electrode cleaning by mechanical or electrochemical procedures, which often affect the electrode stability and reproducibility. The electrode stability and a wider potential window represent the essential features of the electrode for its use in electrolanalysis.

A lot of research studies [2, 9-16] have been reported about the extraordinary properties of boron-doped diamond electrode, e.g., wider potential window for water stability, chemical and mechanical stability, low background currents, resistance to fouling. These properties make this electrode very promising for electroanalysis. However, BDD electrode has the disadvantages of low electrocatalytic activity and selectivity. Thus, the simultaneous detection as overall response of organics responsible for global parameter COD (chemical oxygen demand) that inform about organic loading in water can be achieved [2]. Also, the wider potential window allows to detect simultaneously the individual organics based on the different oxidation potential.

A very interesting aspect in relation with the detection of some organics is given by surface state of the BDD electrode. As-freshed prepared BDD thin film electrode posesses a hydrogen-terminated surface with hydrophobic nature[13], with some non-diamond sp^2-carbon impurities. These impurities may exhibit electrocatalytic activity similarly to graphite electrode, and can be destroyed by anodic oxidation in aqueous electrolytes, when hydrogen-terminated surface is changed to an oxygen-terminated diamond surface with hydrophil character. In general, the presence of high sp^2-non-diamond carbon content lead to the narrow potential window between hydrogen and oxygen evolution in aqueous electrolytes, but some organics can be detect only at hydrogen-terminated surface and not at oxidized surface (e.g., oxalic acid) [14]. After the anodic oxidation as oxygen-terminated state, these can be converted back into hydrogen-terminated one by cathodic treatment in acidic aqueous electrolyte [15].

Chapter 3

CARBON-BASED COMPOSITE ELECTRODES

Carbon-based materials such as graphite, carbon fibers, etc. have largely been used in recent times as the conductive phase in composite materials suitable for electrochemical sensors [17-19]. The use of composites based on a conductive phase dispersed within insulating polymer matrices as electrodes has led to important advances in developing sensor devices, especially in comparison to the single microelectrode systems, in relation with the currents significantly higher obtained due to larger surface areas [20-27]. The composite materials consisted of a combination of two or more materials, in which each individual component exhibit its original nature and should be seen as a whole with the distinctive chemical, mechanical and physical peculiarities [20]. Electrodes obtained by using a mixture of particulate conductive carbon phase and an insulating matrix represent an attractive approach to fabrication of electrochemical sensors, whose surface can be renewed by polishing [17, 21]. The insulating phase has shown growing interest for investigation in order to decrease the electrical resistance and to improve the electrochemical behavior of the composite. In this general context, among composite materials, solid graphite composite electrodes exhibit useful characteristics such as good signal to noise ratio, high mechanical resistance, easy preparation, and renewal and low fabrication cost [28, 29], belong above mentioned good electrical conductivity, chemical inertness, and wide potential range [1, 8].

For some carbon-based composite electrodes, a heterogeneous volume and surface electroactivity with distinct microzones of a significantly higher electrochemical activity have been obtained [22]. The voltammetric responses of carbon-based composite electrodes are sometimes similar to those found for the „edge effect", which contributes significantly to the lowered background current. However, these structured materials are not rigorous arrays but are truly randomised ensembles, with the size, shape and inter-microelectrode separation distributed over a wide range. The random ensembles of microelectrodes can have array behaviour under certain conditions [22]. Microelectrode arrays have been attracting attention for use in electroanalysis because they exhibit the advantages of a macroelectrode which acts as a single microeletrodes, e.g., reduced ohmic drop and charging current, a lower detection limit and better sensitivity. At the same time, their use overcomes the disadvantages of single microelectrode subjected to low-current output and the high susceptibility to the electrochemical noise [29, 30]. In the preparation of carbon-based composites with microelectrode array behaviour, certain important requirements should be met. The main requirement is that the gaps between individual conductive carbon microzones should be much larger than their radii, when spherical diffusion dominates mass transport. The closely spaced carbon microzone arrays will behave similar to a macroelectrode (linear diffusion- controlled mass transport) because of the diffusion layer overlap [31-38]. Another requirement is related to the insulating matrix in order to prevent current leakage, which resulted in the distortion of the cyclic voltammogram [21].

Certain advanced carbon-based composite electrodes, i.e., expanded exfoliated graphite-epoxy (EEGE), carbon nanofiber-expanded graphite-epoxy (CFEGE), expanded graphite-epoxy (EGE), expanded graphite-polystyrene (EGPS), with different contents of conductive phase have recently been prepared and have been characterized structurally and electrochemically, and applied for wet electrochemical detection of the selected organic components [39].

Chapter 4

ZEOLITE-MODIFIED CARBON COMPOSITE ELECTRODES (ZMEs)

The deliberate alteration of the surface of a carbon composite electrode with modified zeolite allows improvement of the electron transfer rate at carbon surface. This type of electrodes form a sub-category of the so-called chemically modified electrodes (CMEs), which meet the needs of many electroanalytical problems. The growing interest for zeolite-modified electrodes arises from the synergistic combination of the attractive properties of the zeolites with electrochemical interfaces. Certain zeolite characteristics are responsible by the possible inprovement for the electron transfer reaction at electrode-solution interphase, such as [40]:

- -the size and shape selectivity due to the zeolite's channels and cages of molecular dimension;
- -cation-exchange capacity arising from charge compensation of the negatively charged aluminosilicate lattice by mobile extra framework cations;
- -catalytic properties of both intrinsic and extrinsic sites of microporous materials.

The applications of zeolite–modified electrodes were described by Walcarius [40], and two major applications are underlined, i.e., electrocatalysis and electroanalysis. The efficiencies of these applications are related to the nature of the zeolite from ZME. The role of this type of electrode in electroanalysis is to improve sensitivity and selectivity for a certain substrate by preconcentration at the electrode-solution interface.

The major contribution of the electrochemical response of ZMEs is ascribed to two components, i.e., extra zeolite electroactive species and electroactive species located within the channels and/or cages at the boundary zeolite-electrode matrix. These components influence the charge-transfer and mass-transfer processes occurring at ZME surfaces. Also, these processes are influenced by the experimental parameters.

A zeolite can be used as support for a catalyst, e.g., copper or silver, in order to improve the performance of electroanalytical detection. Also, aspects of the electroanalysis in relation with the preconcentration and permselectivity can be influenced by the presence of the zeolite. In general, zeolite-modified electrodes exhibit an ion-exchange capacity and a molecular selectivity subjected to size, shape, and charge. Thus, for any electrochemical scheme that relies on ion exchange or analyte preconcentration, and size/shape/charge selectivity or catalyst-assisted reactivity, or any combination of these properties, zeolite-supported electrocatalysts can be used. ZME has been developed for the determination of various compounds, e.g., sugars, herbicides, surfactants, neurotransmitters, and pharmaceuticals [41]. The use of a mixture of a zeolite and graphite in various forms has been proposed [42].

Two types of zeolite-expanded graphite composite electrodes, i.e., Ag – modified zeolite-expanded graphite-epoxy (AgZEGE) and Cu-modified zeolite-expanded graphite-epoxy (CuZEGE) composite electrode have recently prepared, structurally and electrochemically characterized versus expanded graphite-epoxy (EGE) composite electrode, and applied for electrochemical detection of selected molecules [43, 44].

Chapter 5

PREPARATION OF ADVANCED COMPOSITE ELECTRODES MATERIALS [32, 39, 43-47]

Carbon-based composite electrodes are prepared from two-component epoxy resin (LY5052, Araldite) or low-viscosity polystyrene pellets (PS N2000, Crystal from Shell) mixed with two types of conductive carbon fillers, i.e., expanded graphite (EG) powder (Conductograph, SGL Carbon) and carbon nanofibers (CNF) (PS-447 BOX) [39].

The EG filler for the EEGE composite has been subjected in advance to an ultrasonic treatment in an aqueous alcohol solution, following the procedure described in literature [32]. The EEG was then mixed with both epoxy components. The obtained mixture was cured in a hot press at 80 ° C for 40 minutes [39].

For the CFEGE composite, carbon nanofibers and expanded graphite are being mixed with the two components of the epoxy in a roll-mill at room temperature and the obtained paste was cured in a hot press (Fontaijne, Holland) at 80 ° C for 40 minutes. Simultaneousely, the material was shaped in a plate of approximately 1 mm thickness.

In the case of the EGE composite the full amount of EG to the matrix resin was not added directly due to the high surface area of the graphite flakes and, therefore, the mixing was performed in a roll-mill at room temperature. The two parts of the epoxy were mixed together and the full amount of the EG was added in steps forming a thick paste. Then the epoxy was cured in a hot press at 80 ° C for 40 minutes [46].

For all above-described composites, each material was shaped in a plate of 1 mm thickness. The plate was slowly cooled down (for about 12 h) to room temperature without removing the applied pressure.

To prepare the EGPS composite, the graphite filler EG was mixed with the PS N2000 matrix in a roll-mill at 250 ° C for about 30 min. A plate of approximately 1 mm was pressed in a hot press at the same temperature [47].

The zeolite modified electrodes by the types of AgZEGE and CuZEGE are obtained by film casting from two-component epoxy resin (LY5052, Araldite) mixed with conductive expanded graphite (EG) filler powder (Conductograph, SGL Carbon) and silver/copper-doped zeolite (clinoptilolite) [47]. The ratio between the components was chosen to reach 20 weight percent (w/w) content of expanded graphite, 20 weight percent (w/w) content of silver/copper-doped zeolite. It was not possible to add the full amount of EG and Z-Ag to the matrix resin directly, due to the high surface area of the graphite flakes. Therefore, the mixing was performed in a roll-mill at room temperature. The two parts of the epoxy were mixed together and the full amount of EG was added in steps forming a thick paste. Then the epoxy was cured in a hot press at 50 °C for 60 minutes. Simultaneously, the material was shaped in a plate of 1 mm thickness. The plate was cooled down for about 12 h to room temperature [43, 44].

Chapter 6

STRUCTURAL AND ELECTRICAL CHARACTERIZATION OF ADVANCED COMPOSITE ELECTRODES MATERIALS [39, 43 - 49]

SEM images were obtained by using a Scanning Electron Microscope (Philips CM30T). The prepared advanced composites showed good mechanical strength and relative low electrical resistance. The electrical properties of advanced composite electrode materials were determined by the Van der Pauw four-point measurements method [44].

The compositions of the prepared electrodes are gathered in Table 1.

For the carbon-based composite electrode, the plates with a surface area of 9 mm^2 were cut from all compositions and put on glass supports and electrical contacts were made using a silver paint. The electrodes were isolated on the sides by epoxy resin. Prior to use, this working electrode was cleaned, first polished with abrasive paper and then on a felt-polishing pad by using 0.03 μm alumina powder (Metrohm, the Switzerland) in distilled water for 5 minutes and rinsing with distilled water. A picture of EGE electrode is shown in Figure 1a.

Discs with a surface area of 19.63 mm^2 of modified zeolite electrodes were embedded in polyethylene and electrical contacts were made using copper wire. Prior to use, the working electrode was gradually cleaned, first polished with abrasive paper and then on a felt-polishing pad by using 0.3 μm alumina powder (Metrohm, Switzerland) in distilled water for 5 minutes and rinsing with distilled water. An image of this is shown in Figure 1b.

Table 1. The compositions of the carbon-based composite electrodes and Ag/Cu –modified zeolite-expanded grahite-epoxy electrodes

Carbon-based composite electrode type	Insulating matrix type-percent contents (%, wt.)	Conductive matrix type-percent contents (%, wt.)	Zeolite modified with silver or copper*-percent contents (%, wt.)	References
EEGE	Epoxy (95)	Expanded and exfoliated graphite (5)	-	[39]
CFEGE	Epoxy (80)	Carbon nanofiber(10)-expanded graphite (10)	-	[39]
EGE	Epoxy (80)	Expanded graphite (20)	-	[39,45]
EGPS	Polystyrene (80)	Expanded graphite (20)	-	[39, 45, 47]
AgZEGE	Epoxy (40)	Expanded graphite (20)	Ag-modified zeolite (40)	[43]
CuZEGE	Epoxy (40)	Expanded graphite (20)	Cu-modified zeolite (40)	[44]

*8 mg Ag /g zeolite and 5 mg Cu /g zeolite were prepared using natural zeolite from Mirsid, Romania, with 68 weight percent (w/w) clinoptilolite [48].

SEM images provide information about the structure and the distribution of conducting particles in the matrix of epoxy or polystyrene. Also, the surface morphology is examined (Figure 2).

The electrode surfaces are heterogeneous with lighter areas associated with the insulating matrix, and darker areas corresponding to the conducting carbon structures. The carbon flakes are well distributed within both insulating matrices with some differences that can be explained by the different preparation conditions.

Figure 1. Images of advanced composite electrode materials: a) EGE; b) AgZEGE.

A layering of the graphite flakes parallel to the surfaces of the plate is visible for the EGE and EGPS electrodes. For the other composites one can distinguish closely spaced zones of graphite flakes with a random orientation. The higher content of insulating matrix in EEGE suggests a significant surface charging with higher bulk resistance. The morphological differences based on the different natures of the insulating matrices and the carbon flakes orientation and loading led to the different electrical and electrochemical behaviours of the composites electrodes.

Also, for zeolite-modified electrodes, SEM images reveal closely spaced expanded graphite zones with random distribution and orientation due to the irregular shapes of both the expanded graphite particles and Ag or Cu-modified zeolite particles in epoxy matrix.

The electrical properties of advanced composite electrode materials are gathered in Table 2.

Table 2. Electrical properties of composite materials

Carbon composite type	Electrical resistivity ($\Omega\cdot$cm)
EEGE	130
CFEGE	1.43
EGE	0.65
EGPS	0.40
AgZEGE	0.05
CuZEGE	0.08

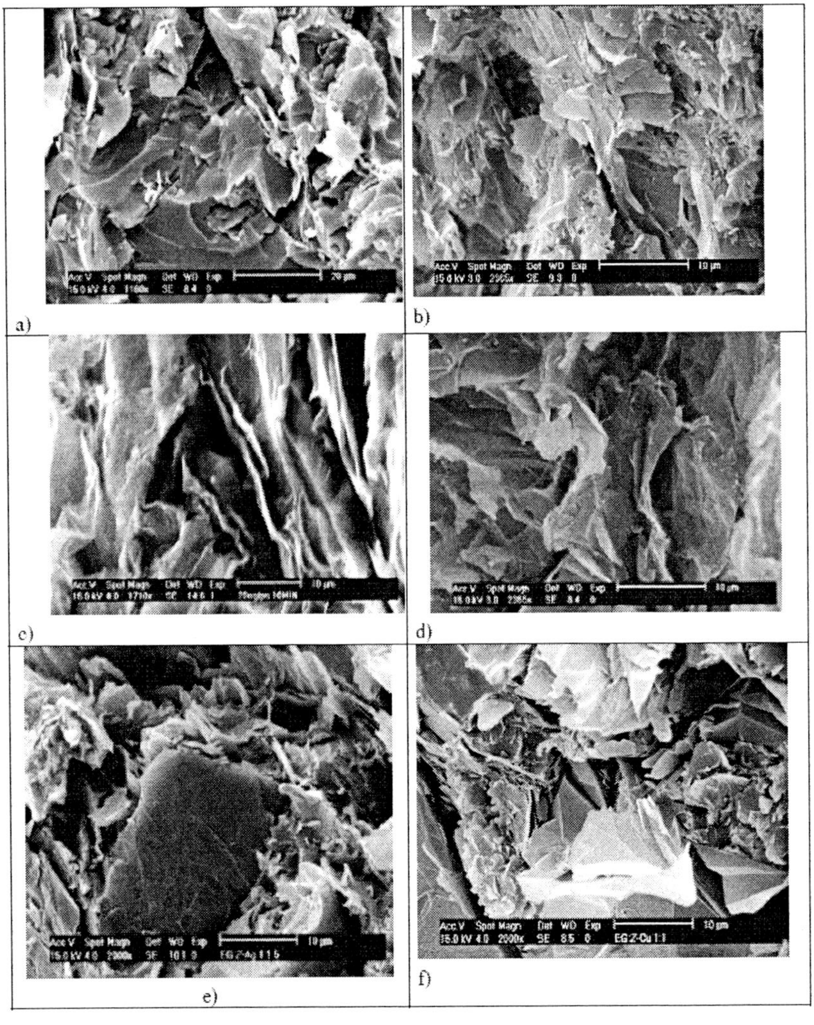

Figure 2. SEM image of a fractured surface of: a) EEGE; b) CFEGE; c) EGE; d) EGPS; e) AgZEGE ; f) CuZEGE.

Chapter 7

ELECTROCHEMICAL BEHAVIOURS AND APPLICATIONS OF ADVANCED CARBON - BASED ELECTRODE FOR ELECTROCHEMICAL DETECTION [39, 43, 44, 46, 47, 49, 50]

The electrochemical behaviours and application performances of the carbon-based composite electrodes in comparison with boron-doped diamond (BDD) electrode were studied by linear-scan voltammetry (LSV), cyclic voltammetry (CV), chronoamperometry (CA), differential pulse voltammetry (DPV), square-wave voltammetry (SWV), and multiple-pulsed voltammetry (MPA) [39, 43, 44, 46, 47, 49-50].

In addition of CV, the electrochemical characterization was studied by Electrochemical Impedance Spectroscopy (EIS). Impedance experiments were carried out with an FRA2 module-containing Autolab PGstat EcoChemie 302 system controlled by a PC, within a frequency range of 65 kHz – 0.1 Hz. The measurements were made continuously at fixed potential values, correlated with cyclic voltammogram. The fitting of EIS measurements was made by Zview program. All potentials were measured versus saturated calomel electrode (SCE) as reference electrode. The impedance spectra were determined at potential values of -0.25; +0.05; +0.35; +0.65; +0.95 and +1.25 V/SCE. These spectra can be analysed with the equivalent circuits (EC) which consisted of: a cell resistance, R1, a constant-phase element (CPE) in parallel with charge transfer resistance, R2.

The different electrochemical pre-treatment in supporting electrolyte for each type of electrode were performed, and the details will be presented in the next section. All measurements were carried out using an Autolab potentiostat/galvanostat PGSTAT 302 (Eco Chemie, The Netherlands) controlled with GPES 4.9 software and a three-electrode cell, with a saturated calomel electrode as reference electrode, a platinum counter electrode, and the composite working electrode.

To improve the sensitivity and lower detection limits for the zeolite modified electrodes, a preconcentration step prior to voltammetric quantification was applied. Accumulation was performed at open circuit potential (OCP) in the analyte solution. The electrode was then removed and rinsed with distilled water, and immersed in the detection cell for recording the voltammetric curve [40]. Regeneration of the working electrode was achieved by mechanical polishing and above-mentioned electrochemical pretreatment.

Chapter 8

ELECTROCHEMICAL BEHAVIOURS OF ADVANCED CARBON-BASED ELECTRODES MATERIALS

CYCLIC VOLTAMMETRY STUDIES

The comparative electrochemical behaviours of BDD and EGE electrodes studied by CV and presented in Figure 3 inform about the potential window of both electrodes in 0.1 M Na_2SO_4 supporting electrolyte. BDD electrode exhibited a larger potential window and a lower background current, aspects very important for electroanalysis.

a)

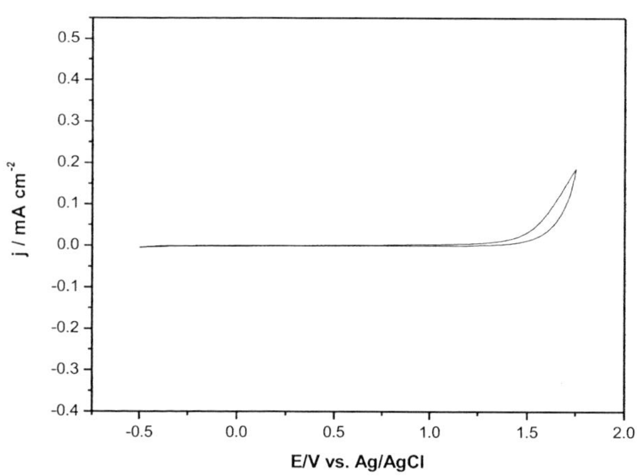

b)

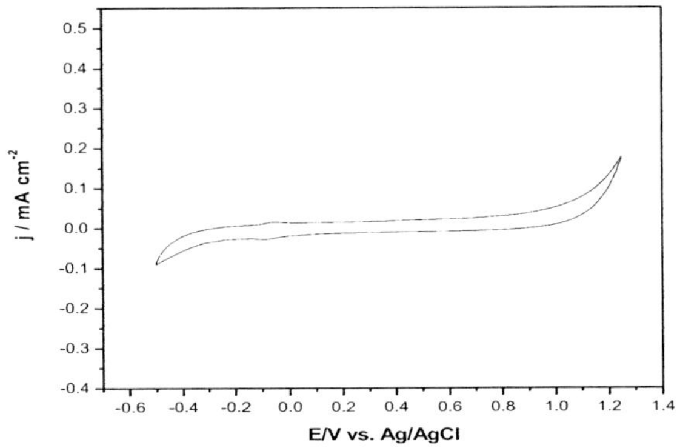

Figure 3. Cyclic voltammograms recorded at a potential scan rate of 0.05 Vs^{-1} in a 0.1 M Na_2SO_4 supporting electrolyte on: a-BDD; b-EGE electrodes.

The electrochemical characterization of graphite-based composite electrodes was conducted by cyclic voltammetry (CV) within the potential range from –0.5 V to +1 V vs. SCE in a 0.1 M Na_2SO_4 supporting electrolyte to obtain comparative information related to the potential window and background current, which are the main important parameters for electroanalysis (Figure 4). An electrochemical pre-treatment by three repetitive cyclings from –0.5 V to +1 V vs. SCE in 0.1 M Na_2SO_4 supporting electrolyte was performed to reach a stable response.

a)

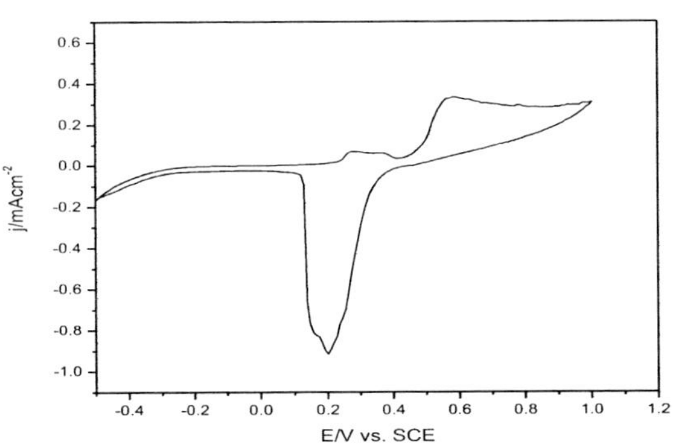

b)

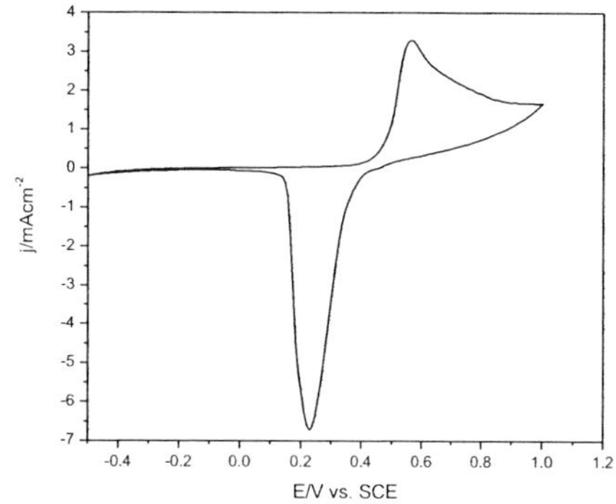

c)

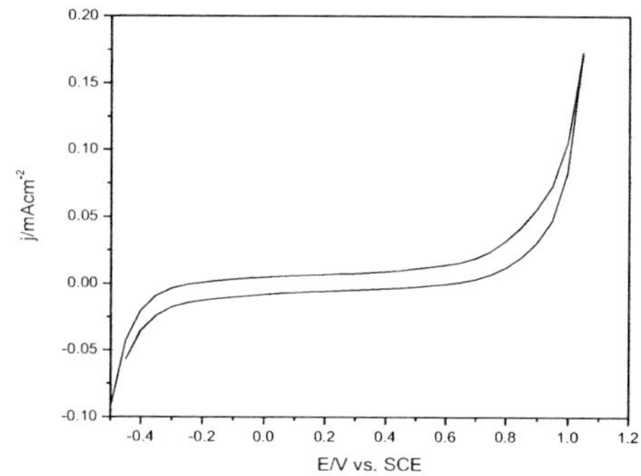

d)

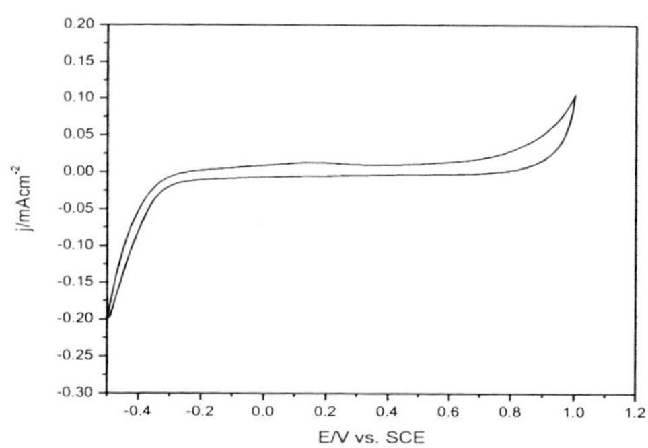

Figure 4. Cyclic voltammograms recorded at a potential scan rate of 0.05 Vs^{-1} between -0.5 and 1V vs. SCE in a 0.1M Na_2SO_4 supporting electrolyte on: a-EEGE; b-CFEGE; c-EGE; d-EGPS electrodes.

The EEGE and CFEGE electrodes exhibited a well-defined anodic current peak, noticed on the CVs around 0.6 V vs. SCE, with a corresponding cathodic one (around 0.2 V/SCE), owing to the functional groups containing graphite electrode material [14, 50].The current density of the CFEGE electrode is much higher than the one registered an EEGE electrode, related to the different capacitive behaviours of the two electrodes, function of electrode composition. The electrochemical behaviours of EGE and EGPS electrodes are different in comparison with previous electrodes, no functional groups of graphite were manifested [49]. The potential window was very slightly influenced by the composition and the morphology of the carbon- based electrodes.

a)

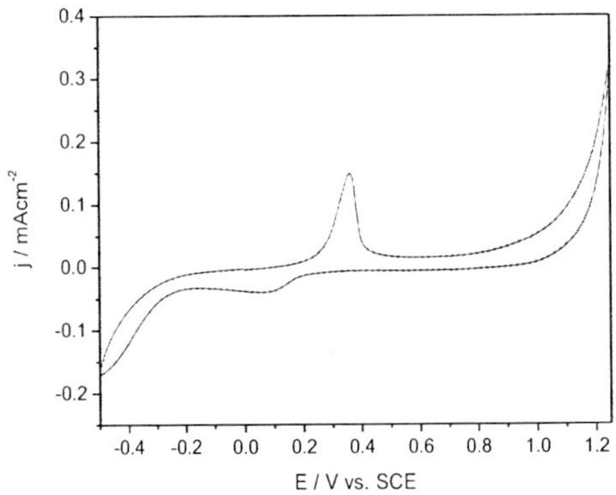

b)

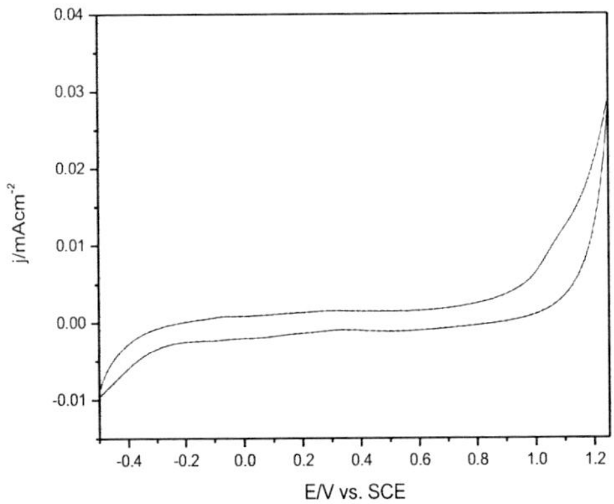

c)

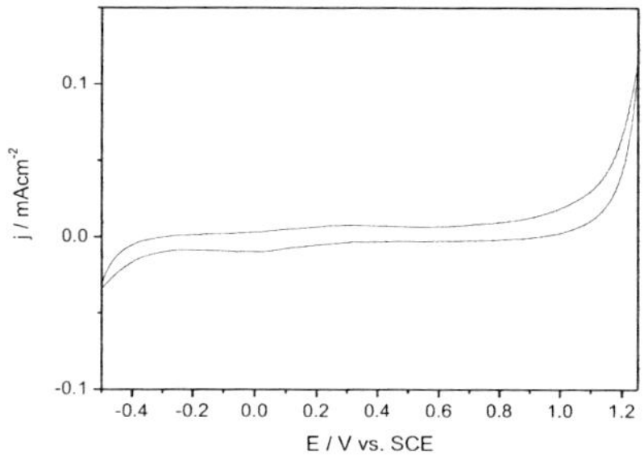

Figure 5. Cyclic voltammograms recorded at a potential scan rate of 0.05 Vs^{-1} between -0.5V and 1.25 vs. SCE in 0.1M Na_2SO_4 supporting electrolyte for AgZEGE (a), CuZEGE(b) and EGE electrode (c).

a)

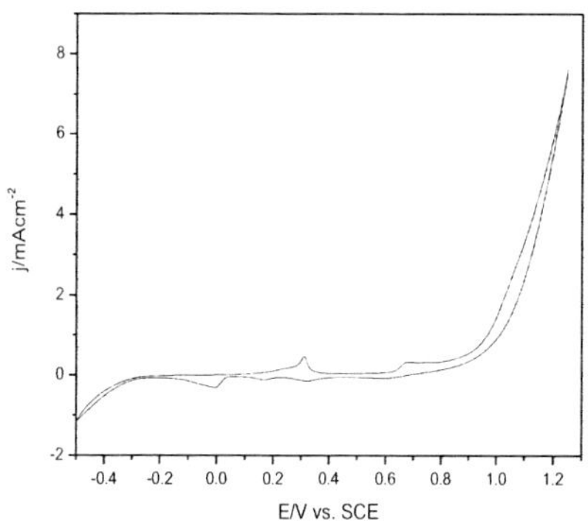

b)

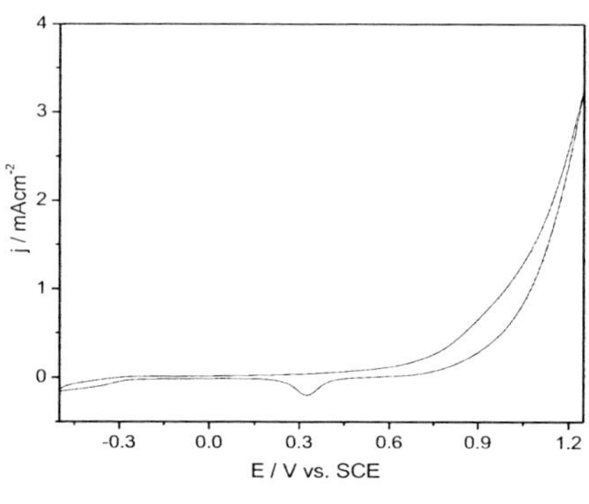

c)

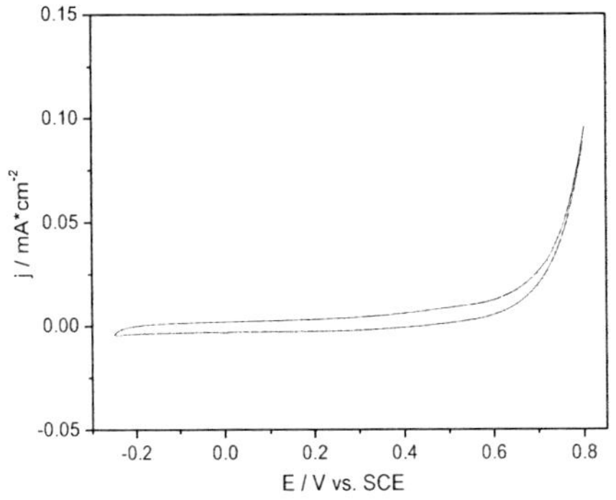

Figure 6. Cyclic voltammograms recorded at a potential scan rate of 0.05 Vs^{-1} between -0.5V and 1.25 vs. SCE in 0.1M NaOH supporting electrolyte for AgZEGE (a), CuZEGE (b) and EGE electrode (c).

The electrochemical characterization of the AgZEGE electrode was performed in comparison with the EGE electrode by CV within the potential range from –0.5 V to +1.25 V vs. SCE in 0.1 M Na_2SO_4 supporting electrolyte.

Figure 5 presents the cyclic voltammograms (CVs) of AgZEGE electrode in comparison with CVs of EGE in a 0.1 M Na_2SO_4 supporting electrolyte. As expected, the AgZEGE electrode exhibited the oxidative/reductive peak corresponding to the redox peaks of the Ag/Ag(I) couple. For this electrode, the background current is higher than for the EGE electrode, also a common aspect for the electrocatalytic electrode type.

Figure 6 depicts CVs of AgZEGE, CuZEGE, and EGE composite electrodes in 0.1 M NaOH supporting electrolyte. It can be seen that the expanded graphite composite electrodes containing Romanian zeolite with small amount of Ag or Cu cations, distributed randomly in the zeolite structure, display complex behaviour with several electrochemically distinct species.

For the AgZEGE electrode (Figure 6a), the anodic part of the voltammogram is characterized by the occurrence of the anodic peaks corresponding to the electroformation of soluble $[Ag(OH)_2]^-$ complex species (the first peak, shoulder), to the electroformation of Ag_2O (the second peak), and to the formation of AgO (the third peak) [51-53]. The cathodic part of the cyclic voltammograms is characterized by the occurrence of cathodic peaks corresponding to the electroreduction of AgO to Ag_2O, and the reduction of Ag^+ located in different sites of zeolite [40].

The modifications of the expanded graphite-epoxy matrix with Z-Cu provided an active electrode surface in alkaline solutions, without the presence of peaks/shoulder corresponding to copper oxidation [54], may be due to the small electrocatalyst amount and the inhomogeneous distribution within the composite material. However, the voltamograms (Figure 6 b) showed a reduction peak at around 0.32V/SCE due to the electrochemical reduction of Cu^{2+} to Cu^+ [55]. The background current as obtained only in 0.1 M NaOH was much higher for the ZME electrode than for the EGE electrode, mainly due to double layer charging. Also, since the charging currents are proportional to the real area rather than the geometric one, it is likely that ZME electrodes are characterized by a more porous morphology.

An investigation of the effect of the scan rate on the peak currents and potentials, involving O_2 and H_2 evolution was carried out and a very complex behaviour of the electrodes was noticed (Figure 7). A thorough investigation related to mechanism and kinetics parameters was carried on the redox AgO/Ag_2O couple. The results of the scan rate effect on this couple, which exhibits interest from the electrocatalytic point of view, are shown in Figure 8a.

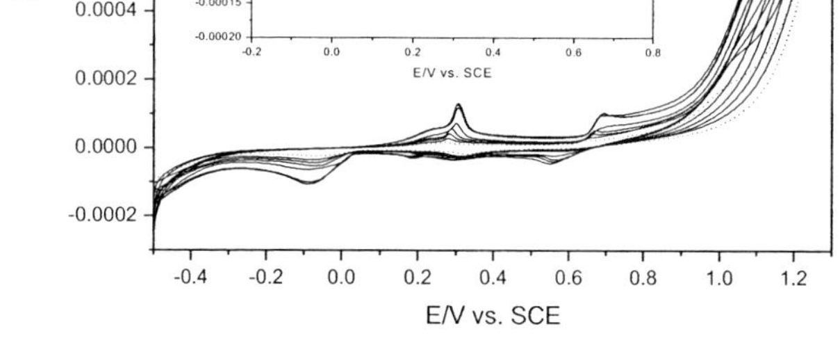

Figure 7. Cyclic voltammograms (CVs) of AgZEGE electrode in 0.1 M NaOH supporting electrolyte at different scan rates: 1-0.01, 2-0.02, 3-0.03, 4-0.04, 5-0.06, 6-0.08, 7-0.1 Vs^{-1}, from -0.5 V to +1.25V vs. SCE; Inset: Detail of a limited range (-0.3 V to +0.8 V vs. SCE).

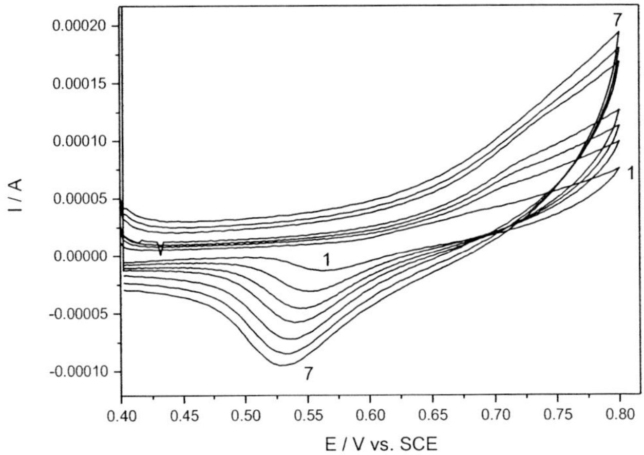

Figure 8a. Cyclic voltammograms (CVs) of an AgZEGE electrode in 0.1 M NaOH supporting electrolyte at different scan rates: 1-0.01, 2-0.02, 3-0.03, 4-0.04, 5-0.06, 6-0.08, 7-0.1 $V \cdot s^{-1}$ from +0.4 V to +0.8 V vs. SCE.

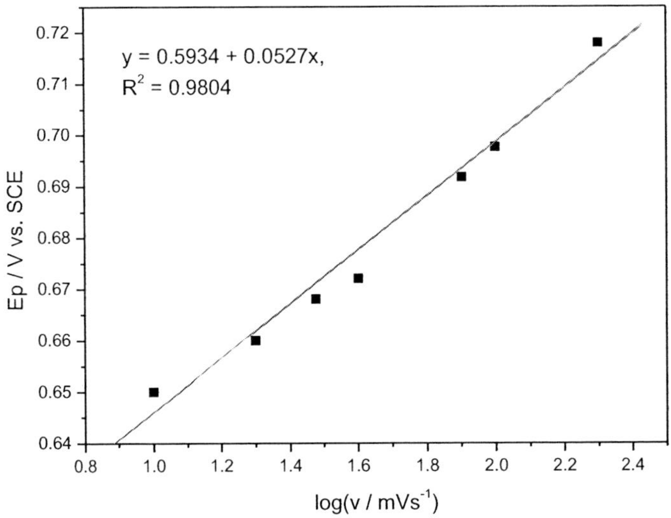

Figure 8b. Dependence of the anodic peak potential versus the logarithm of the scan rate.

The linear proportionality of the current peak vs. square root of the scan rate (the results are not shown here) indicated that the process was generally diffusion -step controlled and the practically zero intercepts suggested that adsorption steps and surface interactions were negligible. Also, from the variation of peak potentials with scan rate, using the treatment proposed by Laviron [56], kinetic parameters for the heterogeneous electron transfer at AgZEGE composite electrode in relation with redox couple of Ag(II) and Ag(I) oxides were determined. A plot of Ep=f(logv) yields two straight lines with slopes equal to 2.3RT/αnF and 2.3 RT/(1-α)nF for the cathodic and anodic peak (see Figure 8b, as example for the anodic peak). Based on these plots and the equation [51]:

$$\log k_s = \alpha \log(1-\alpha) + (1-\alpha)\log \alpha - \log(RT/nFv) - \alpha(1-\alpha)nF\Delta Ep/2.3RT \quad (1)$$

the values of the transfer coefficient (α) and the apparent charge-transfer rate constant (k_s) were 0.42 and 0.36 s^{-1}, respectively, for EG-Z-Ag-Epoxy in the presence of 0.1 M NaOH subjected to AgO /Ag_2O couple.

Chapter 9

ELECTROCHEMICAL IMPEDANCE SPECTROSCOPY STUDIES

Further electrochemical characterization of BDD, EGE, AgZEGE, CuZEGE electrodes by electrochemical impedance spectroscopy technique (EIS) allow to obtain information about the reaction mechanism and electrode kinetics at the electrode-electrolyte interface[57]. EIS should be considered a complementary technique of CV, which allow examining the electrodes after initial characterization of potential window by CV. Based on CV results and taking into account that the values of potential to cover the whole potential window, the EIS experiments were carried out in 0.1 M Na_2SO_4 and 0.1 M NaOH supporting electrolyte at values of applied potential of -0.25; +0.05; +0.35; +0.65; +0.95; +1.25 V (vs. SCE). Based on CV results related to the potential window, BDD electrode was characterized in addition at +1.55 V/SCE.

To analyze comparatively the data for all electrodes one single simple equivalent circuit (EC) can be used for fitting, which consisted of: a cell resistance, R1, a constant-phase element (CPE) in parallel with charge transfer resistance, R2 (Figure 9). The CPE is assumed to be a non-ideal capacitance with roughness factor α, with α equal to 1 for a perfectly smoothed surface area and 0.5 for a porous electrode [57]. The experimental Nyquist diagrams for tested electrodes are shown in Figures 10 a-d and Figures 11 a-c. Fitting was carried out with ZView 2.3 software.

The values of EC parameters obtained from fitting are gathered in Table 3.

In 0.1 M Na_2SO_4 supporting electrolyte, the cell resistance R1 varied in according with the electrical resistivity of each electrode. BDD electrode exhibited the highest electrical resistance.

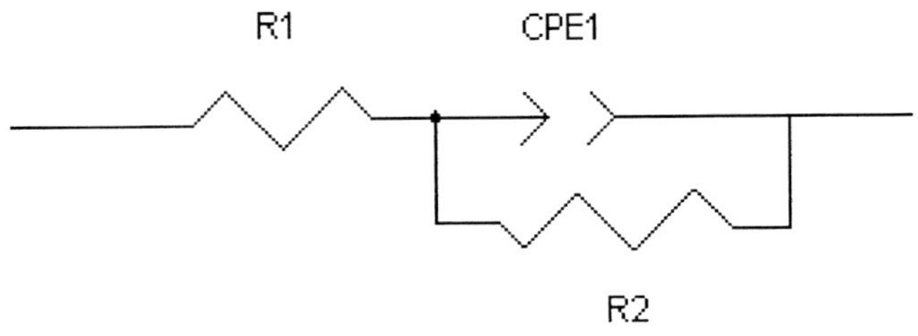

Figure 9. Equivalent circuit used for fitting the impedance spectra at carbon-based electrodes.

For the same electrode, R1 varied a little according to the supporting electrolyte. The capacitance values for each supporting electrolyte are lowest for BDD electrode related to a less compact double layer. The most compact double layer was obtained for Ag and Cu zeolite modified- expanded graphite composite electrode. The higher capacitance and respective, higher double layer corresponding to higher background current are characteristics to the electrode with catalytic properties. There is no significant difference in the roughness factor, α. The values of α higher than 0.7 suggest the smooth electrode surface. The lowest R1 values at the anodic extreme potential (+1.25 and respective, +1.55 vs. SCE for BDD) are related to the oxygen evolution. For Ag /Cu ZEGE electrodes, the R1 values are much smaller because of some electrode reactions occured, demonstrated by semi-circular form of the spectra.

a)

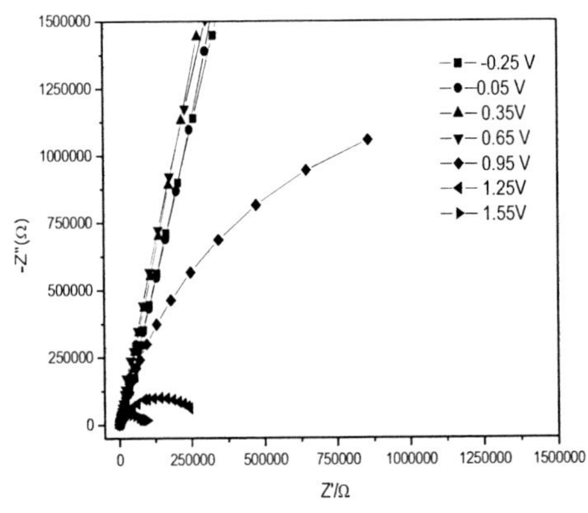

b)

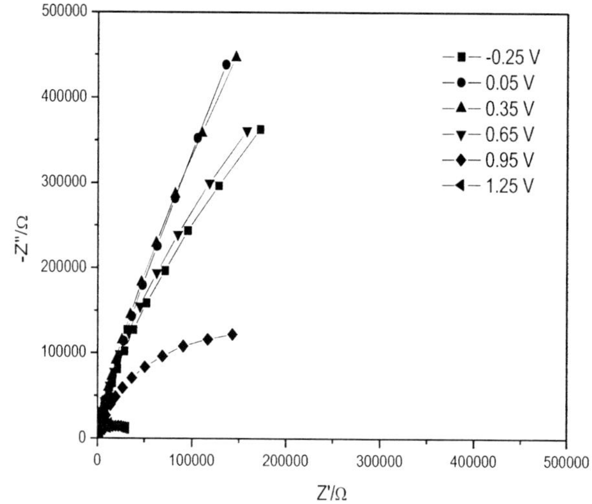
500000
400000
300000
200000
100000
0
-Z''/Ω
0
100000
200000
300000
400000
500000
Z'/Ω
-0.25 V
0.05 V
0.35 V
0.65 V
0.95 V
1.25 V

c)

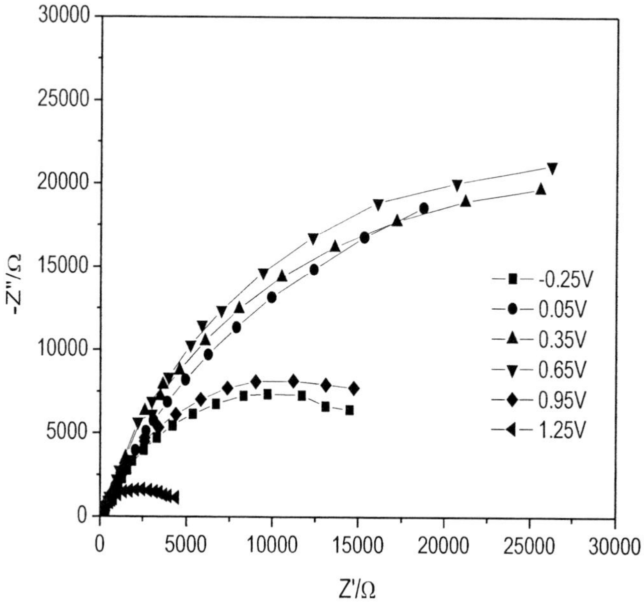
30000
25000
20000
15000
10000
5000
0
-Z''/Ω
0
5000
10000
15000
20000
25000
30000
Z'/Ω
-0.25V
0.05V
0.35V
0.65V
0.95V
1.25V

d)

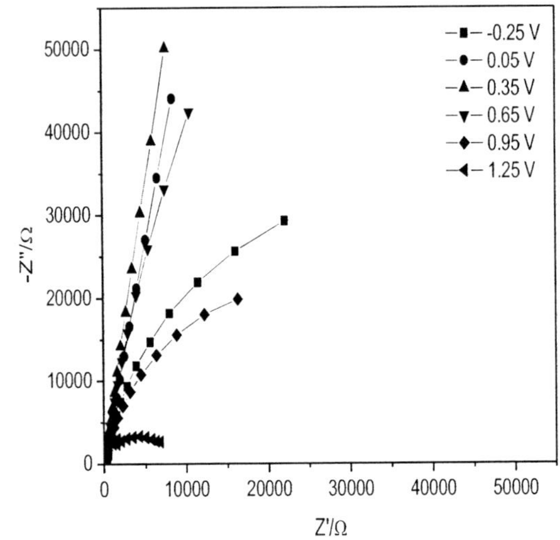

Figure 10. The comparative experimental Nyquist diagrams obtained in 0.1 M Na2SO4 supporting electrolyte at each potential value for the: BDD(a) EGE (b); AgZEGE (c); CuZEGE (d).

a)

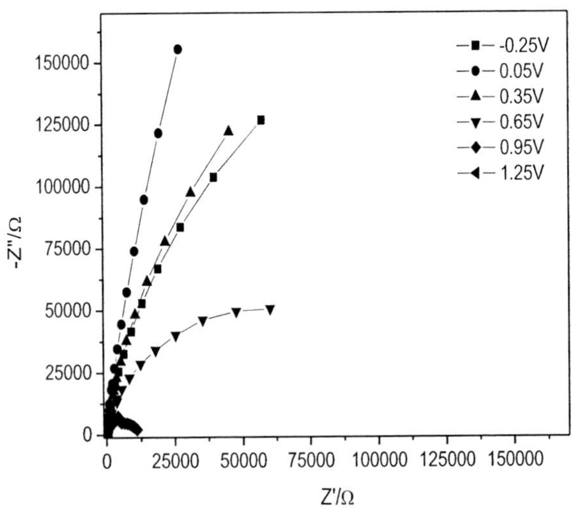

b)

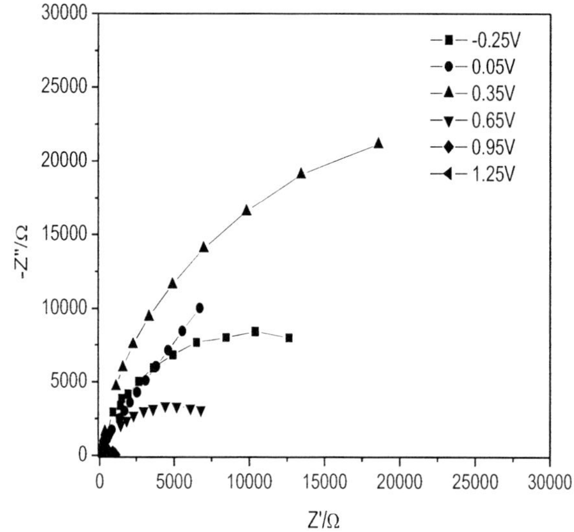

c)

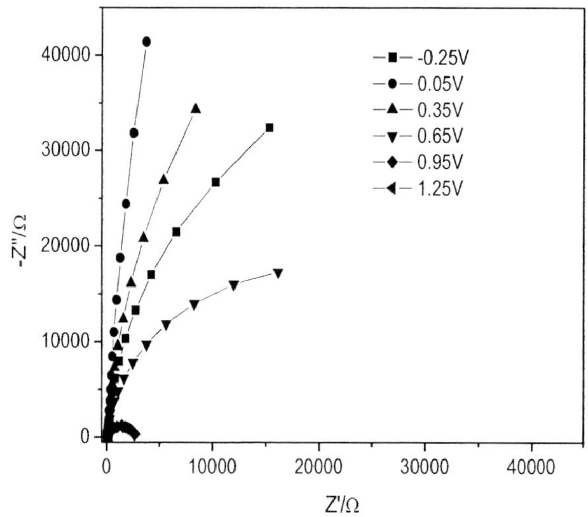

Figure 11. The comparative experimental Nyquist diagrams obtained in 0.1 M NaOH supporting electrolyte at each potential value for the: a) EGE, b) AgZEGE; c) CuZEGE electrodes.

Table 3. Data obtained from analysis of the impedance spectra for carbon-based electrodes

Electrode	E (V/SCE)	R1 (Ω cm^{-2})	C1 (μF cm^{-2})	α	R2 (kΩ cm^{-2})
0.1 M Na_2SO_4 supporting electrolyte					
	-0.25	1867.6	3.59	0.872	$0.35 \cdot 10^6$
	0.05	1863.3	10.11	0.877	$0.31 \cdot 10^6$
	0.35	1905.8	11.10	0.882	$0.76 \cdot 10^6$
	0.65	1934.1	10.52	0.890	$0.58 \cdot 10^6$
BDD	0.95	1868.9	12.42	0.879	$0.38 \cdot 10^5$
	1.25	1785.5	15.40	0.847	$0.36 \cdot 10^4$
	1.55	1915.7	4.24	1	$0.17 \cdot 10^4$
EGE	-0.25	936.6	17.6	0.964	$0.89 \cdot 10^4$
	0.05	944.1	29.8	0.880	$0.45 \cdot 10^5$
	0.35	911.9	29.0	0.880	$0.22 \cdot 10^5$
	0.65	907.8	33.0	0.880	$0.12 \cdot 10^5$
	0.95	891.7	41.7	0.890	$0.24 \cdot 10^5$
	1.25	898.4	92.6	0.870	$0.39 \cdot 10^4$
AgZEGE	-0.25	412.7	187.0	0.757	$0.10 \cdot 10^3$
	0.05	410.0	191.1	0.746	$0.28 \cdot 10^3$
	0.35	426.8	42.8	0.970	$0.15 \cdot 10^3$
	0.65	426.2	54.5	0.950	$0.16 \cdot 10^3$
	0.95	433.9	115.7	0.880	$0.92 \cdot 10^2$
	1.25	434.6	387.3	0.770	$0.25 \cdot 10^2$
CuZEGE	-0.25	539.6	171.2	0.880	$0.41 \cdot 10^3$
	0.05	523.8	162.5	0.870	-
	0.35	545.2	171.2	0.880	-
	0.65	546.8	168.7	0.890	$0.33 \cdot 10^4$
	0.95	541.1	208.9	0.880	$0.24 \cdot 10^3$
	1.25	569.7	205.9	0.950	$0.37 \cdot 10^2$
0.1 M NaOH supporting electrolyte					
EGE	-0.25	868.3	107.8	0.935	$0.46 \cdot 10^4$
	0.05	867.6	103.3	0.939	$0.20 \cdot 10^5$
	0.35	869.3	118.9	0.936	$0.52 \cdot 10^4$
	0.65	877.4	140.0	0.939	$0.12 \cdot 10^4$
	0.95	884.4	108.4	0.990	$0.13 \cdot 10^3$
	1.25	871.6	245.6	0.845	$0.15 \cdot 10^1$
AgZEGE	-0.25	356.6	188.5	0.915	$0.97 \cdot 10^3$
	0.05	354.2	319.4	0.825	$0.23 \cdot 10^3$
	0.35	364.0	78.7	0.850	$0.29 \cdot 10^3$
	0.65	375.8	146.2	0.870	$0.35 \cdot 10^2$
	0.95	389.3	400.5	0.862	$0.49 \cdot 10^1$

Electrode	E (V/SCE)	R1 (Ω cm^{-2})	C1 (μF cm^{-2})	α	R2 (kΩ cm^{-2})
	1.25	388.5	671.6	0.720	0.40
CuZEGE	-0.25	486.5	186.0	0.965	$0.46\cdot10^3$
	0.05	493.0	184.7	0.971	$0.46\cdot10^4$
	0.35	493.2	204.4	0.968	$0.90\cdot10^3$
	0.65	495.9	229.4	0.972	$0.18\cdot10^3$
	0.95	495.7	321.3	0.939	$0.13\cdot10^2$
	1.25	495.6	272.1	0.902	0.30

Chapter 10

ELECTROCHEMICAL CHARACTERIZATION AND APPLICATION OF ADVANCED COMPOSITE ELECTRODES MATERIALS IN RELATION WITH MICROELECTRODE ARRAY BEHAVIOUR

Previous investigations have demonstrated that the carbon-based composite electrode can be regarded either as an ordered (array) or randomized microelectrode ensembles [17, 22, 31]. Under these circumstances, the diffusion flux toward microelectrode ensembles depend strongly on the experimental timescale. Thus, a linear diffusion towards each individual microelectrode occurred for the short timescale, and the spherical diffusion field dominates microelectrode ensemble responses for the intermediate timescale. At sufficiently long time, the diffusion layers totally overlap giving a macroelectrode behavior in which the signal is proportional to the whole microelectrode ensembles as sum of the electroactive and isolate areas [31-36].

It is known that a microelectrode array has a dramatically lower capacitive current than a conventional macroelectrode. The capacitive current, which is proportional to the active electrode area, scan rate and capacitance is the main cause of the background current in voltammetry. Reducing the capacitive current can greatly improve the signal to noise ratio. In Table 4 are presented the useful signals expressed as the difference between the peak current in the presence of 1mM thiourea (TU) chosen as electroactive species, and the background current.

Table 4. The useful responses of advanced composite electrode materials in 0.1 M Na_2SO_4 supporting electrolyte containing 1 mM TU

Type of advanced composite electrodes materials	Useful signal for 1mM TU/mA
EEGE	0.934
CFEGE	0.955
EGE	6.39
EGPS	6.54

Because the possibility to observe microelectrode array behavioral aspects depends on the size, geometry and the gaps between microelectrodes, but also on the response time of the electrode, the scan rate in the cyclic voltammetry was therefore systematically varied over a wide range of scan rates from 0.001 to $0.05 V \cdot s^{-1}$ and Figures 12 a-d illustrate the effects of the scan rate on the cyclic voltammograms of each electrode material for 1 mM TU containing 0.1 M Na_2SO_4 supporting electrolyte. For CFEGE composite electrode, the current responses increase linearly with increasing scan rate, while for the other composite electrodes the scan rate effect is slightly, expressed as the square root of the scan rate (Figure 13). However, for EEGE composite electrode, a significant hystersis between the forward and backward scans is noticed. For EGE and EGPS composites electrodes, a current plateau has been observed when the scan rate was lower than approximately $0.05 V \cdot s^{-1}$. The small effect of the scan rate and a current plateau suggest that a pseudo-steady state response has been achieved with little hysteresis between the forward and backward branches of the cyclic voltammograms. The limiting current is independent of the scan rate. According to the results, the EGE and EGPS composite electrode materials are considered to be in a random array [58].

The ability of the microelectrode array to deliver current responses in the absence of any supporting electrolyte is an attractive feature if compared with the conventional macroelectrodes [37]. There are several reasons subjected to analytical studies on nondilute samples in the absence of any supporting electrolyte, *i.e.*, the direct analysis in low ionic strength solutions that make analysis more facile, eliminating the possible interference of the supporting electrolyte and the possibility of extending the upper concentration limit of the analyte under investigation.

a)

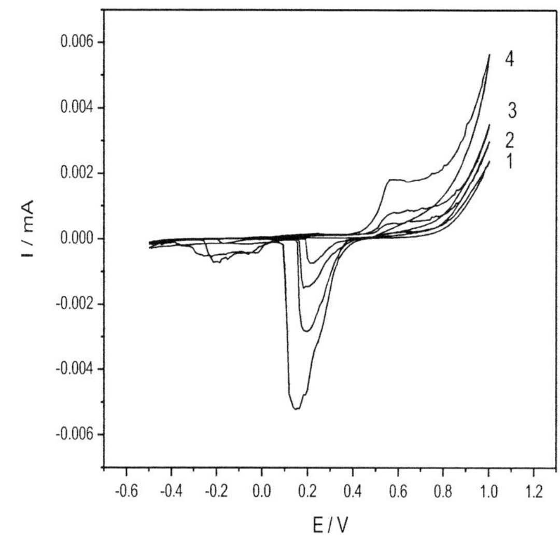
0.006
0.004
0.002
0.000
-0.002
-0.004
-0.006
I / mA
-0.6 -0.4 -0.2 0.0 0.2 0.4 0.6 0.8 1.0 1.2
E / V
4
3
2
1

b)

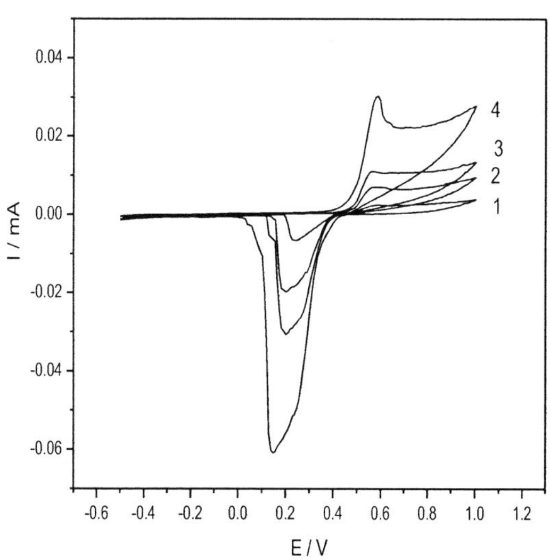
0.04
0.02
0.00
-0.02
-0.04
-0.06
I / mA
-0.6 -0.4 -0.2 0.0 0.2 0.4 0.6 0.8 1.0 1.2
E / V
4
3
2
1

c)

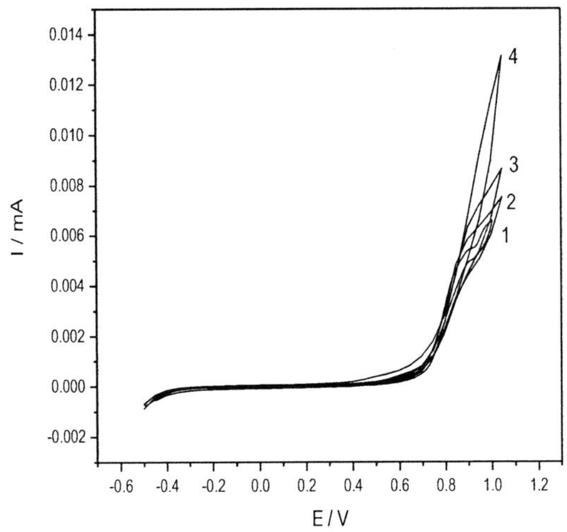

d)

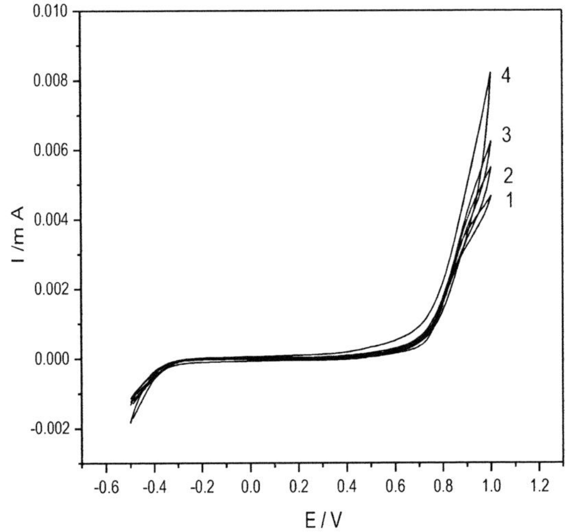

Figures 12 a-d. Cyclic voltammograms in 1mM TU solution and 0.1 M Na_2SO_4 supporting electrolyte at different scan rates: 1-0.001 $V \cdot s^{-1}$, 2-0.005 $V \cdot s^{-1}$, 3-0.01 $V \cdot s^{-1}$, 4-0.05 $V \cdot s^{-1}$ at: a)EEGE, b) CFEGE, c) EGE , d) EGPS electrodes.

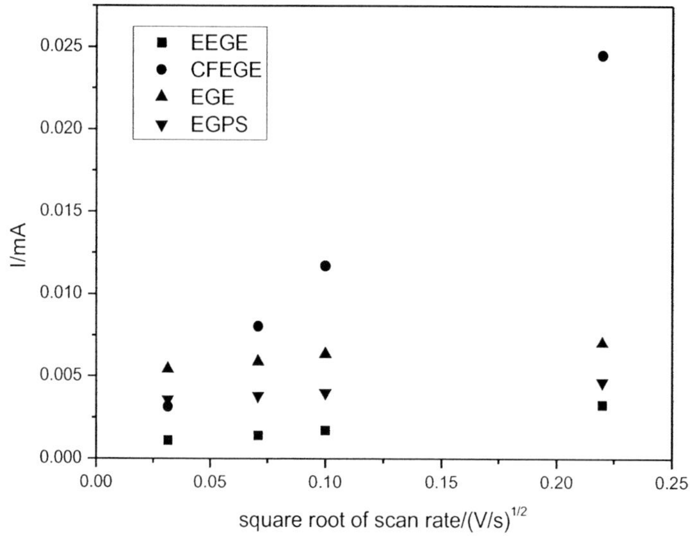

Figure 13. Dependence of voltammetric current recorded at +0.9V vs. SCE in 1mM TU solution and 0.1 M Na_2SO_4 supporting electrolyte versus the square root of the scan rate for each advanced composite electrode material.

Based on the electrochemical behavior of the studied carbon-based composite electrode, the EGE electrode was tested for TU detection in tap water, under the conditions of not deliberately added supporting electrolyte, using LSV, CV, and CA. The scan rate of 0.05 $V \cdot s^{-1}$ was chosen to keep the ohmic drop negligible. In Figure 14 is shown the linear-scan voltammograms recorded in tap water and in the presence of different TU concentrations, with the linear dependence between anodic current at +0.9 V/ SCE and TU concentration, which is presented in Table 5.

The cyclic-voltammogram responses obtained at EGE under the same conditions of LSV application are shown in Figure 15. The anodic current recorded at +0.9 V vs. SCE increases with increasing of TU in tap water and a relative low hysteresis was observed. Inset of Figure 11 shows that the useful anodic current responses from forward branches of CVs recorded at +0.9 V/SCE depend linearly on the TU concentration in the range 0.1-1.0 mM with the corresponding parameters gathered in Table 5.

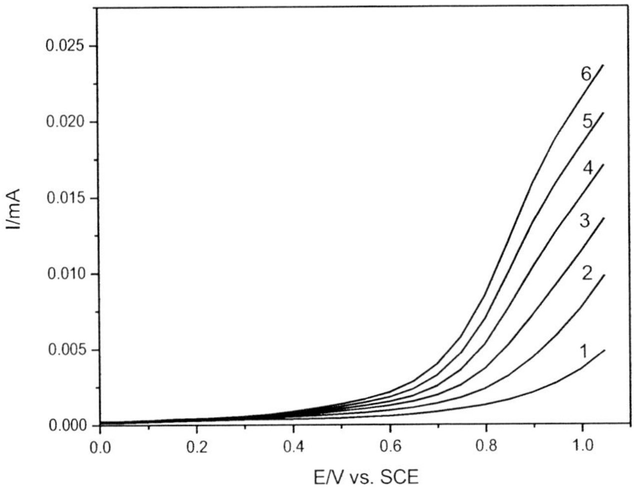

Figure 14. Linear-scan voltammograms of EGE electrode in tap water and in the presence of different TU concentrations:1-0 mM, 2-0.2 mM, 3-0.4 mM, 4-0.6 mM, 5-0.8 mM, 6-1 mM; potential scan rate of 0.05 Vs^{-1}.

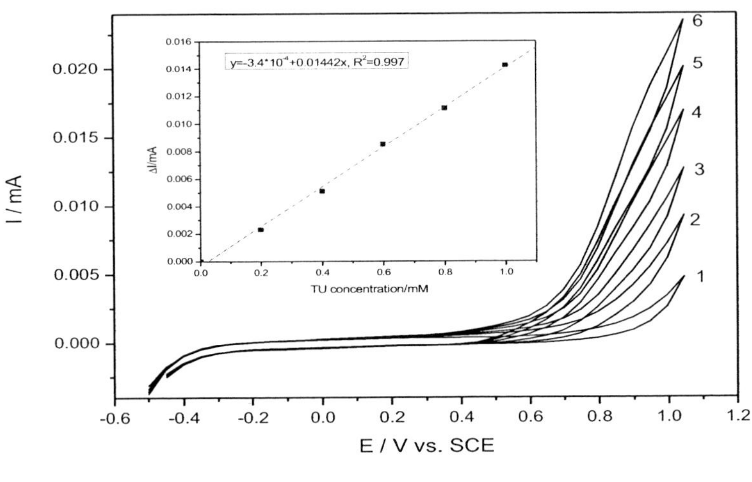

Figure 15. Cyclic voltammograms of EGE in tap water in the presence of different TU concentrations:1-0 mM; 2-0.2 mM; 3-0.4 mM; 4-0.6 mM; 5-0.8 mM; 6-1 mM; potential scan rate 0.05 Vs^{-1}. Inset: the useful anodic current from forward branches of CVs recorded at 0.9 V/SCE vs. TU concentrations.

An example of hydrodynamic chronoamperograms (Figure 16) was obtained by successively adding TU to a quiescent tap water. The applied potential was set at +0.9 V vs. SCE and the response time was about 50 seconds.

A linear dependence of current versus TU concentration was obtained in the concentration range of 0.5 to 2.5 mM with the correlation coefficient of 0.98. For further TU adding at the concentration higher than 2.5 mM TU, a response platform was observed, probably because of some remanent adsorption effects on the surface of the electrode, leading to electrode fouling. This effect should also be responsible for the lower sensitivity of this chronoamperometric assessment of 0.00463 $mA \cdot mM^{-1}$, compared to LSV and CV analytical data (Table 5).

Under these experimental conditions, no cleaning or pretreatment of the electrode after each addition of TU was required. The analytical parameters are presented in Table 4. The lowest limit of detection (LOD), was evaluated based on $3S_B/b$ [59], where S_B is the standard deviation of the mean value of 5 voltammograms (amperograms) of the blank and b is the slope of the straight line in the analytical curve by using the respective technique. The reproducibility of the data at the EGE electrode using the above-mentioned techniques was evaluated for three replicates measurements of TU detection. The relative standard deviations (RSD) gathered in Table 5 show the good reproducibility of the electrode.

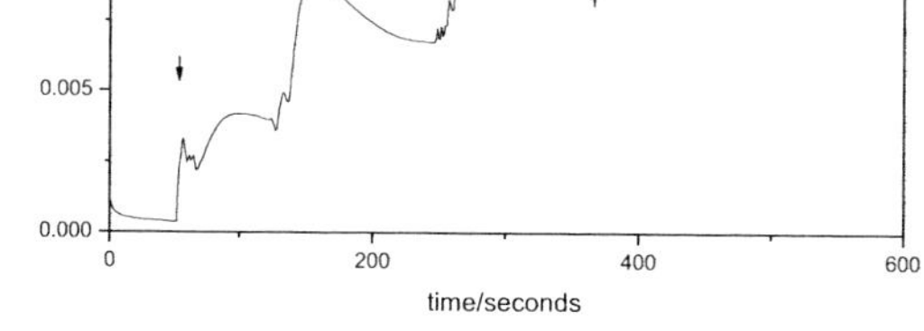

Figure 16. Chronoamperometric records obtained at EGE for successive additions of 0.5 mM TU in tap water without supporting electrolyte; working potential:+0.9 V/SCE.

Table 5. The analytical parameters of voltammetric/amperometric detection of TU at an EGE electrode using different techniques

Technique used	Concentration range(mM)	Electrode sensitivity (mA·mM^{-1})	Correlation coefficient (R^2)	LOD (mM)	RSD (%)
LSV	0.1-1	0.01443	0.9980	0.03	2.7
CV	0.1-1	0.01442	0.9992	0.02	2.5
CA	0.5-2.5	0.00463	0.9832	0.10	3.2

Chapter 11

ELECTROANALYTICAL APPLICATION OF BDD ELECTRODE

Boron-doped diamond electrode, due to its electrochemical properties desired for electroanalysis, high stability and wider potential window have been used for many electroanalytical application [9-16, 59]. In this sub-chapter the electroanalytical parameters for the detection of individual phenol derivates, oxalic acid and their simultaneous detection are presented.

INDIVIDUAL DETECTION OF ORGANICS AT BDD ELECTRODE

Individual Detection of Phenol Derivates

Phenol derivates are toxic and non-biodegradable organics, which are released into the environment, being commonly used/produced in several industrial processes, e.g., preservatives, desinfectants, in pulp processing, in the manufacturing of pesticides [60-65]. Due to their toxicity to humans and most aquatic organisms and their tendency to bioaccumulate in the food chain [66], the European Union [67] and the US Environmental Protection Agency [68] have labelled chlorinated phenols as "priority pollutants". Many priority phenols are known for their toxicity, carcinogenetic and environmental persistence character. They are common pollutants found in water, sediments, and soil. Based on the toxicity, stability and bioaccumulation, the European Union has set a „black list" of the dangerous substances that should be monitored in waters. The phenol derivates were included in that list, which was later accepted in the Water Framework Directive [69].

Among the different phenol derivates pollutants of aquatic ecosystems, some examples of the electrochemical detection of each compounds at BDD electrode are shown further in relation with mono and di-substituted phenol derivates, e.g., 4-chlorophenol (4-CP), 4-aminophenol (4-AP), and 2,4-dinitrophenol (2,4-DNP).

The severe toxicity of 4-CP requires the development of a simple, sensitive and reliable determination method. Different chromatographic [70-73] and spectrometric techniques [74] have been used in the past for the detection of chlorophenols in wastewater. However, these methods require expensive and most often laboratory complicated instruments. The specific cost of analysis can be high, and the methods are not suitable for in-situ measurements. Electrochemical techniques for 4-CP oxidation and determination has attracted a growing attention in the past decade, due to the low costs, fast analysis, and direct applicability of electrochemical sensing to a wide range of analytes often in particular hostile media [39, 43, 75-82].

Some aspects regarding 4-CP oxidation and determination of 4-CP at BDD are studied by CV and differential pulsed voltammetry (DPV). DPV can be considered as a derivative of linear-scan voltammetry, with a series of regular voltage pulses superimposed on the potential linear sweep, which are dependent on applied parameters, *i.e.*, a scan rate of 25 mVs^{-1}, a pulse modification of 50 mV in amplitude and 50 ms in duration at intervals of 200 ms.

Cyclic voltammograms (CVs) obtained at BDD electrode in the presence of 0.2 and 1 mM 4-CP in 0.1 M Na_2SO_4 are shown in Figure 17. The current corresponding to the potential value of about +0.5 V vs. Ag/AgCl owing to the direct oxidation of 4-CP, increased with its concentration.

The cyclic voltammetry of BDD electrode at various scan rates (0.01-0.2 $V{\cdot}s^{-1}$) in the presence of 0.2 mM 4-CP was studied (Figure 18). For the whole range of the scan rates studied, the peak shape that is sensitive to the scan rate shows irreversible characteristics.The anodic current densities at about +0.6 V vs. Ag/AgCl increased linearly with the square root of the scan rate (inset *a* of Figure 18) suggesting that the reaction is mass-transfer controlled. Even the potential value of the anodic peak remain constantly with increasing v, which are caracteristics to reversible process (inset *b* of Figure 18), on the following reverse scan from +1.75 to -0.5 V vs. Ag/AgCl, no corresponding reduction peak is observed, revealing that the electrode process on the BDD is totally irreversible.

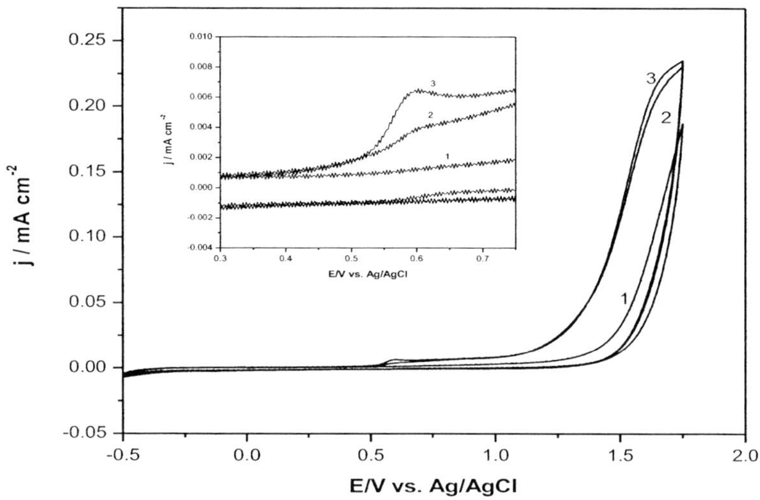

Figure 17. First scan of cyclic voltammogram of BDD electrode in 0.1 M Na_2SO_4 supporting electrolyte (curve 1), in the presence of 0.2 mM 4-CP (curve 2), in the presence of 1 mM 4-CP (curve 3) ; potential range: $-0.5V \rightarrow +1.75V \rightarrow -0.5V$ vs. Ag/AgCl; scan rate of 0.05 Vs^{-1}. Inset: detail of limited potential range.

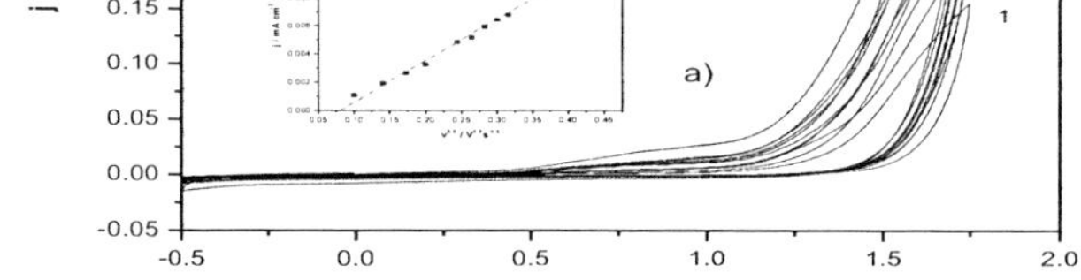

Figure 18. Cyclic voltammograms of BDD electrode in 0.1 M Na_2SO_4 supporting electrolyte and in the presence of 0.2 mM 4-CP at the different scan rate: curve 1-0.01V s^{-1}; curve 2-0.02 Vs^{-1}; curve 3-0.03 V s^{-1}; curve 4-0.04 Vs^{-1}; curve 5-0.05 Vs^{-1}; curve 6-0.06 Vs^{-1}; curve 7-0.07 Vs^{-1}; curve 8-0.08 V s^{-1}; curve 9-0.09 Vs^{-1}; curve 10-0.1 $V{\cdot}s^{-1}$;curve 11-0.2 V·s-1; potential range: $-0.5V \rightarrow +1.75V \rightarrow -0.5V$ vs. Ag/AgCl. Insets: a) – Dependence of current densities of anodic peaks registered at 0.6 V vs. Ag/AgCl function of $v^{1/2}$, b) - Dependence of the potential of current peak, Ep to the log (v).

In Figures 19 and 20, linear-scan voltammograms and differential-pulsed voltammograms in the presence of different 4-CP concentrations are shown. The dependences of the current peaks vs. concentration, explored in the concentration range between 0.5 mM-4 mM, were linear, with the very good correlation coefficients, 0.997 and respective, 0.995. The sensitivity of 0.002 $mAcm^{-2}mM^{-1}$ and respective, 0.003 $mAcm^{-2}mM^{-1}$ was reached. Only the above-presented working conditions for DPV applying a very sligt sensitivity enhancement was reached.

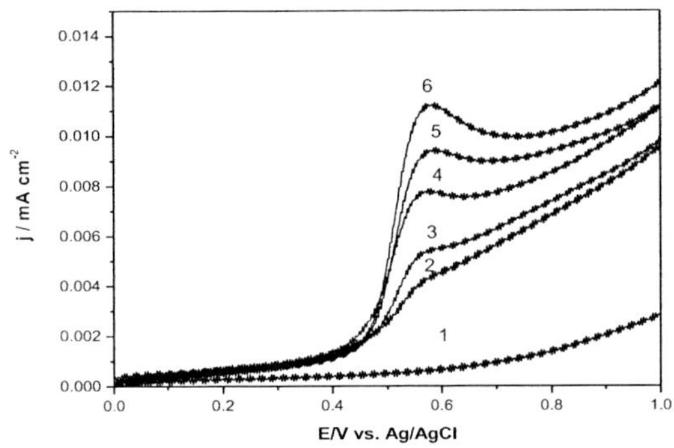

Figure 19. Linear-scan voltammograms of BDD electrode in the 0.1 M Na_2SO_4 supporting electrolyte (curve 1) and in the presence of various concentrations of 4-CP: 1-0 mM; 2-0.5 mM; 3-1mM; 4-2 mM; 5-3 mM; 6-4 mM; potential scan rate: 0.05 Vs^{-1}.

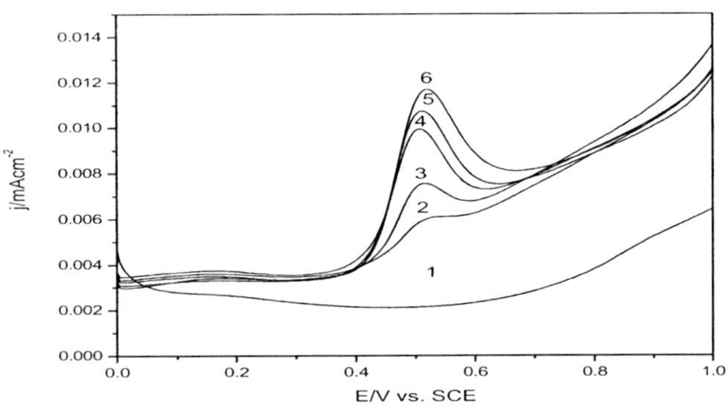

Figure 20. Differential-pulsed voltammograms of BDD electrode in the 0.1 M Na_2SO_4 supporting electrolyte (curve 1) and in the presence of various concentrations of 4-CP:1-0 mM; 2-0.5 mM; 3-1mM; 4-2 mM; 5-3 mM; 6-4mM; potential scan rate: 0.05 $Vs.^{-1}$.

4-Aminophenol (4-AP) constitutes an important category of organic water pollutants, regarded mainly as intermediate products of degradation process of some industrial compounds, e.g., dyes, pesticides and industrial chemicals (nitrobenzene, acetanilide) [83, 84]. Also, 4-AP is a metabolic product of the insecticide parathion and aniline pollution in the body and is excreted in urine as glucurono-and sulfo-conjugates. 4-AP is found during the storage of pharmaceutical preparations or during the synthesis of paracetamol [85, 86]. All these aspects require a fast method for 4-AP determination from both environmental and biological point of view.

Figure 21 shows the cyclic voltammograms obtained at BDD electrode in different 4-AP concentrations. Compared with 4-CP, from the shape of voltammogram can be seen that 4-AP is more easily oxidizable, the anodic limiting current appearing at the potential value of 0.07 V vs Ag/AgCl. On the reverse scanning, the cathodic peak corresponding to the reduction process appeared at about -0.19 V vs. Ag/AgCl. The potential difference between anodic and cathodic peak higher than 59 mV and the ratio between anodic and cathodic peaks different by 1 suggested a quasi-reversible redox process. Based on the linear dependence with a good correlation coefficient of 0.995 between anodic peak current recorded at +0.07 V vs. Ag/AgCl and 4-AP concentration, the sensitivity of 0.269 $mAcm^{-2}mM^{-1}$for the voltammetric determination of 4-AP at BDD was determined.

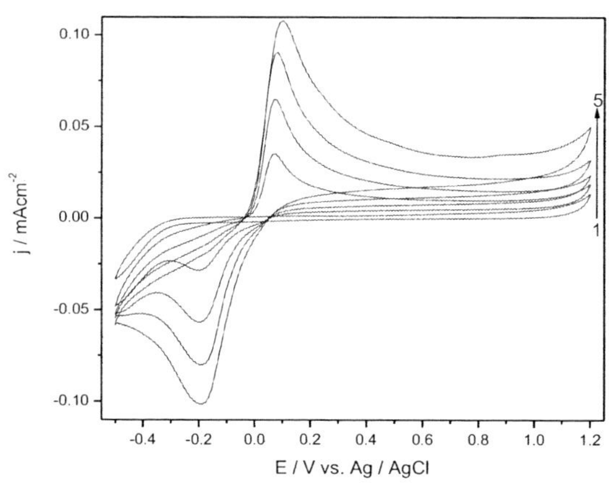

Figure 21. Cyclic voltammograms of BDD electrode in the 0.1 M Na_2SO_4 supporting electrolyte (curve 1) and in the presence of various concentrations of 4-AP:1-0 mM; 2-0.1 mM; 3-0.2 mM; 4-0.3 mM; 5-0.4 mM; potential scan rate 0.05 Vs^{-1}.

The effect of the potential scan rate on the CV recorded on BDD in 0.1 M Na_2SO_4 and 0.2 mM 4-AP is shown in Figure 22. No electrode cleaning during CV running at different scan rates was performed.

The current density of anodic peak registered at +0.07 V vs. Ag/AgCl is not proportional with the square root of the scan rate (inset of Figure 22), which indicates that the oxidation process of 4-AP by direct electron-transfer reaction is not controlled by mass transport, and could involve a surface process. It is known that phenol derivates are oxidised to form a radical cation that can be further oxidized and/or it can react to form an electrically insulating polyoxyphenylene film that adhere to the electrode surface [79, 82]. Increasing the potential scan rate at value higher than 0.07 Vs^{-1}changed the forward anodic branch by appearance of two additional anodic peaks at more positive potentials, about + 0.5 and +0.9 V vs. Ag/AgCl corresponding to further oxidation of 4-AP.

In comparison with the effect of the scan rate on BDD electrode in the presence of 4-CP, the sorbtion and/or surface processes are more evident for 4-AP. This could be explained by the higher rate of the formation of the polymer film on the side, and on the other side, by the limited potential ranges, which correspond to the water stability. In this range, no oxygen evolution occured, and no mineralization was achieved.

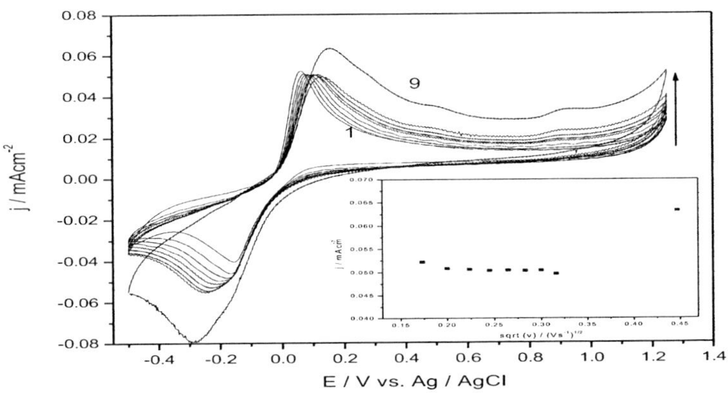

Figure 22. Cyclic voltammograms of BDD electrode in 0.1 M Na_2SO_4 supporting electrolyte and in the presence of 0.2 mM 4-AP at the different scan rate: curve 1-0.03V s^{-1}; curve 2-0.04 Vs^{-1}; curve 3-0.05 V s^{-1}; curve 4-0.06 Vs^{-1}; curve 5-0.07 Vs^{-1}; curve 6-0.08 Vs^{-1}; curve 7-0.09 Vs^{-1}; curba 8-0.1 V s^{-1}; curve 9-0.2 Vs^{-1}; potential range: $-0.5V \rightarrow +1.25V \rightarrow -0.5V$ vs. Ag/AgCl.

In Figure 23 is shown an example of hydrodynamic chronoamperograms recorded at BDD electrode by successively adding 0.05 mM 4-AP to 0.1 M Na_2SO_4 supporting electrolyte. The applied potential was set at +0.1 V vs. Ag/AgCl and the response time was about 50 seconds.

A linear dependence of current versus 4-AP concentration was obtained in the concentration range of 0.05 to 0.35 mM with the correlation coefficient of 0.98. Under these conditions, the sensitivity of 0.038 $mAcm^{-2}mM^{-1}$was achieved, which was much lower than one determined by CV, which can be own to the electrode fouling effect.

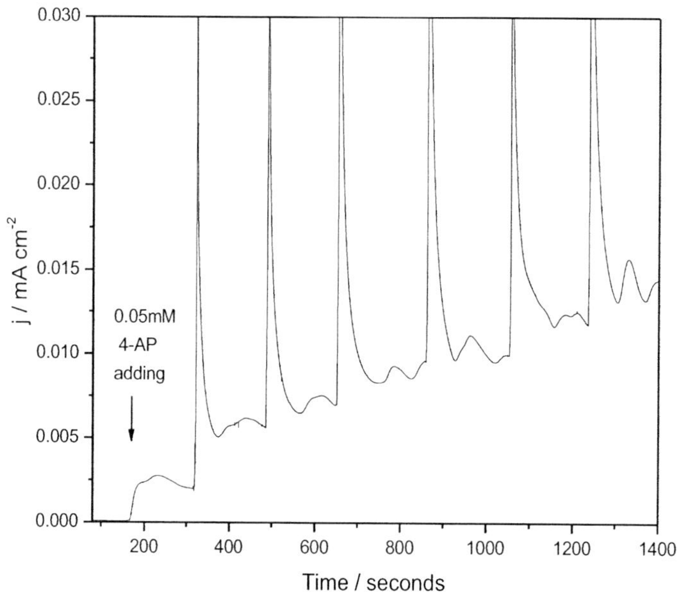

Figure 23. Chronoamperometric records obtained at BDD electrode for successive additions of 0.05 mM 4-AP in 0.1 M Na_2SO_4 supporting electrolyte; working potential: +0.1 V vs. Ag/AgCl.

Another phenol derivate studied for its electrochemical determination on BDD electrode was 2,4-dinitrophenol (2,4-DNP). Some aspects related to the electrochemical behaviour of 2,4-DNP were given by Canizares et al. [87] in order to treat the aqueous wastes containing 2,4-DNP. The anodic peaks appeared in the range of phenol oxidation (1.3 and 1.6 V vs. Ag/AgCl), and no corresponding reduction cathodic peaks appeared (Figure 24). Though, with

2,4-DNP concentration increasing, it was noticed an increasing of the cathodic current recorded at about -0.5 V vs. Ag/AgCl corresponding to the reduction of nitrophenols to aminophenols derivates [82]. The peaks decrease in size with the number of cycles, because of the electrode fouling.

To overcome the fouling problem, which appeared during the electrode is used in a fixed potential amperometry mode (chronoamperometry), cleaning of electrode surface can be carried out by mechanical polishing or electropolishing. Electropolishing is more convenient because it can be employed at any time during analysis. Multiple pulsed amperometry (MPA) is a type of electropolishing, which utilizes a potential waveform to generate new electroactive electrode surface prior to each current measurement. Unlike common amperometric measurement, the applied potential waveform is an alternating potential between positive and negative values [88].

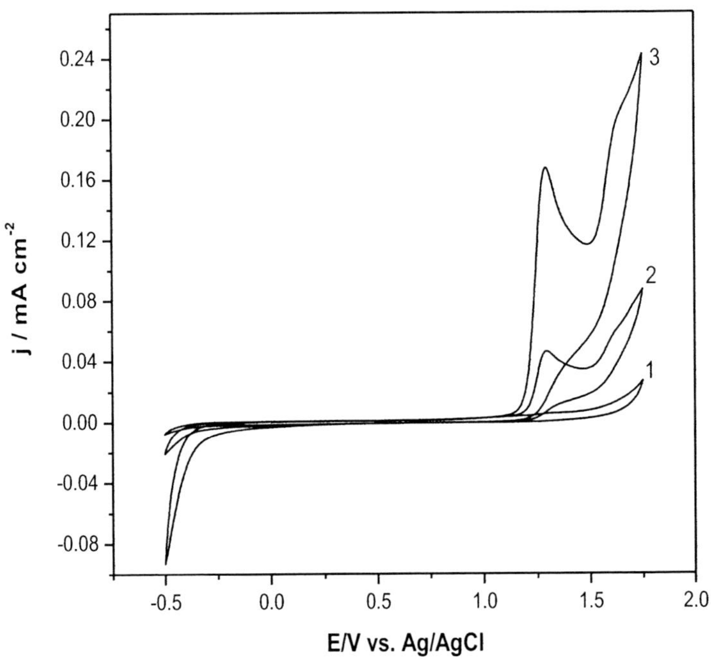

Figure 24. Cyclic voltammograms in 0.1 M Na_2SO_4 (curve 1) and in the presence of 0.2 mM (curve 2) and 1mM (curve 3) 2,4-DNP at BDD electrode; potential range: -0.5 V→ +1.75 V → -0.5V vs. Ag/AgCl; scan rate: 0.05 Vs^{-1}.

Under the conditions of MPA applying (E_{det} = + 1.3 V, t_{det} = 0.05 s and E_{clean} = -0.5 V, t_{clean} = 0.15 s), an improvement in terms of higher sensitivity was reached [89]. For the electrochemical determination of 2,4-DNP, it was determined a sensitivity of 0.28 mA $cm^{-2}mM^{-1}$ by using MPA (Figure 25) versus 0.03 mA $cm^{-2}mM^{-1}$ by CA applying.

BDD electrode showed different electrochemical activity for the electrochemical oxidation and determination of various types of phenol derivates, i.e., 4-chlorophenol, 4-aminophenol, and 2,-dinitrophenol. The electroanalytical parameters for their electrochemical determination at BDD electrode are gathered in Table 6.

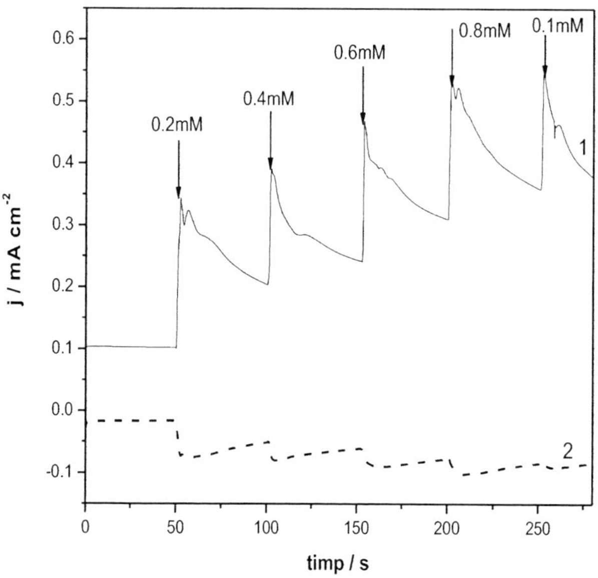

Figure 25. Multiple-pulsed amperogram (MPA) recorded at BDD electrode in 0.1 M Na_2SO_4 supporting electrolyte and in the presence of different 2,4-DNP concentrations; E_{det} was +1.3 V (curve 1) and E_{clean} was -0.5 V vs. Ag/AgCl (curve 2); t_{det} was 0.05 seconds and tclean was 0.15 seconds.

Table 6. The electroanalytical parameters for the electrochemical determination of 4-CP, 4-AP, and 2,4-DNP at BDD electrode

Phenol derivate type	Concentration range(mM)	Electrochemical technique	Oxidation potential	Sensitivity (mA·mM^{-1})
4-CP	0.5-4	LSV	+0.5 vs. SCE	0.002
	0.5-4	DPV	+0.5 vs. SCE	0.003
4-AP	0.1-0.4	CV	+0.07 vs. Ag/AgCl	0.269
	0.05-0.35	CA	+0.1 V vs. Ag/AgCl	0.038
2,4-DNP	0.2-1	CA	+1.3 V vs. Ag/AgCl	0.030
	0.2-1	MPA*	+1.3 V vs. Ag/AgCl	0.280

*cleaning potential of -0.5 V vs. Ag/AgCl.

Electrochemical Individual Detection of Oxalic Acid (OA)

Oxalic acid naturally occurs in various organisms, including fungi, plants and animals. High levels of oxalic acid remove calcium from blood with severe disturbances in the activity of heart and neural system. It is also known that oxalic acid may cause digestive tract irritation and kidney damage, and increasing demand for its determination in urine appeared [90-92]. Therefore, oxalic acid is a metabolite of the advanced oxidation of phenol and its derivates, which are present in many industrial wastewaters and other aromatic substrates of pesticides, herbicide, and fungicides. Some simple compounds, for instance oxalic acid is known as recalcitrant compound and resistant to further degradation [93-95]. A number of methods for the determination of oxalic acid, for example, the titration method [92], the fluorimetric method [96], the spectrophotometric [97], the HPLC [98-102], the chemiluminescence [103], the HPLC- chemiluminescence [104], the AAS [91], the enzymatic method [105], and the electrochemical detection [14] were reported. Most of the methods require a preliminary separation of oxalic acid from its matrix, or lack of the sensitivity, efficiency or simplicity. The electrochemistry of oxalic acid at some metal electrodes [95, 106, 107], Pd-modified GC [94] and BDD [14] has been studied for its removal or determination.

In comparison with the results related to oxalic acid determination at BDD electrode reported by Ivandini et al.[14], the oxidation potential peak corresponding to the oxidation of oxalic acid is shifted at more positive potential value (about +1.6 V vs. SCE compared with +1.32 V vs. Ag/AgCl). Figure 26 shows a series of LSVs for the oxalic acid oxidation at various concentrations (0.5-4 mM) in 0.1 M Na_2SO_4 supporting electrolyte. In Figure 27 is shown the results of DPV applied for the oxalic acid detection. The sensitivity obtained by both methods are similar, about 0.042 $mAcm^{-2}mM^{-1}$.

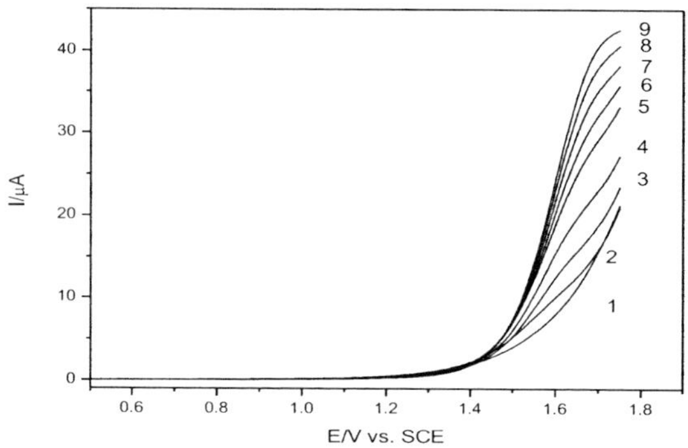

Figure 26. Linear-scan voltammograms of the BDD electrode in the 0.1 M Na_2SO_4 supporting electrolyte (curve 1) and in the presence of various concentrations of OA: 0.5 mM to 4 mM (curves 2-9), concentration step: 0.5 mM OA; potential scan rate 0.05 Vs^{-1}.

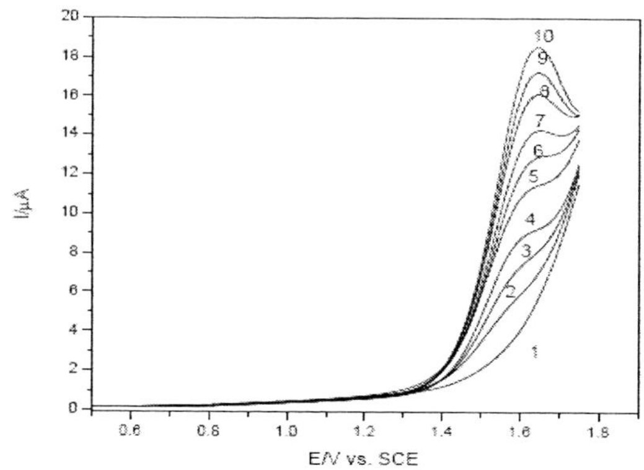

Figure 27. Differential-pulsed voltammograms of BDD electrode in the 0.1 M Na_2SO_4 supporting electrolyte (curve 1) and in the presence of various concentrations of OA: 0.5 mM to 4.5 mM (curves 2-10), concentration step: 0.5 mM OA; potential scan rate 0.025 V s^{-1}.

Electrochemical Simultaneous Detection of Oxalic Acid (OA) and 4-Chlorophenol (4-CP)

In general, chlorophenols and saturated carboxylic acids are refractory compounds to conventional chemical and biological oxidation [107]. The oxygenated compounds, such as saturated carboxylic acid, are the intermediates, or the end products of the conventional and advanced oxidation process of chlorophenols contained in waste waters [108-112].

Under these conditions, the simultaneous determination of chlorophenol and saturated carboxylic acid in water is required especially to elucidate the mechanism of oxidation process. The presence of saturated carboxylic acid, e.g., oxalic acid (OA) in water after the application of the oxidation process of 4-CP shows that no complete mineralization occurred. The determination of saturated carboxylic acid as intermediary or end product for the 4-CP oxidation by using HPLC [113] and electrochemical methods [39, 50, 62] was reported.

BDD electrode was tested to detect OA in the presence of 4-CP. Based on the electrochemical behaviour of BDD electrode in the presence of each individual organic compound (discussed in the above section), the difference between oxidation peak potentials of OA (1.5 V/SCE) and 4-CP (0.5 V/SCE) should allow the detect OA in the presence of 4-CP.

Figure 28 shows an example of linear-scan voltammograms for 4-CP and the mixture of OA and 4-CP at BDD electrode. The presence of OA did not influence the potential peak corresponding to 4-CP oxidation and detection. Also, the presence of OA was detected without significant intereference of 4-CP.

Using DPV and CA, OA was detected in the presence of 4-CP with no requirement of complicated procedures. 4-CP was not influenced significantly the sensitivity of OA detection on BDD electrode (Table 7). In Figure 29 is shown a series of DPVs at various concentration of OA (0.5-3 mM) and in the presence of 1 mM 4-CP.

Also, the linear relationship of peak currents versus OA concentrations were obtained in the absence/presence of 4-CP, and the electrode sensitivity for OA detection was almost the same either in the absence and in the presence of 4-CP. Figure 30 shows a comparative example of chronoamperograms in the absence and in the presence of 4-CP recorded at +1.65 V/SCE during succesive adding of 0.5 mM OA in 0.1 M Na_2SO_4 supporting electrolyte.

Table 7. Comparative electroanalytical parameters for OA detection at BDD electrode in the presence/absence of 4-CP

Concentration range	Electrochemical technique	Sensitivity/$mAcm^{-2}mM^{-1}$	
		No presence of 4-CP	Presence of 4-CP
0.5-4	CA	0.017	0.020
0.5-4.5	DPV	0.042	0.043

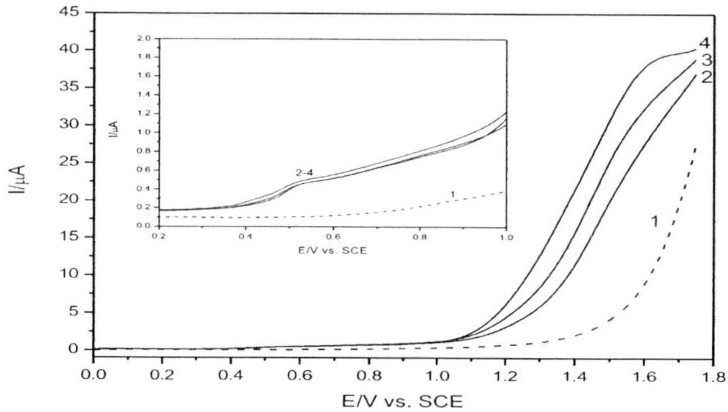

Figure 28. Linear-scan voltammograms of BDD electrode in a 0.1 M Na_2SO_4 supporting electrolyte (curve 1), in the presence of 0.5 mM 4-CP(curve 2), in the presence of 0.5 mM 4-CP and 0.5 mM OA (curve 3), in the presence of 0.5 mM 4-CP and 1 mM OA (curve 4); Inset: detail for the limited potential range.

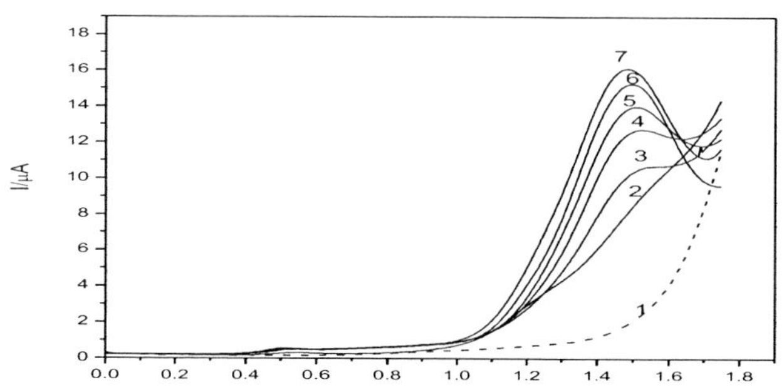

Figure 29. Differential- pulsed voltammograms of BDD electrode in the 0.1 M Na_2SO_4 supporting electrolyte (curve 1), and in the presence of 1 mM 4-CP (curve 2), and in the presence of various concentrations of OA: 0.5 mM to 3 mM (curves 3-7), concentration step: 0.5 mM OA; potential scan rate 0.05 Vs^{-1}.

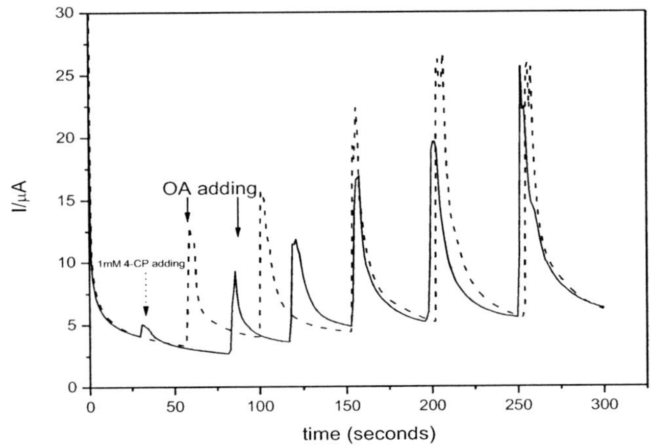

Figure 30. Chronoamperograms of BDD electrode in 0.1 M Na_2SO_4 supporting electrolyte (dash line) and the presence of 1 mM 4-CP (solid line) during continuous adding of OA; concentration range:0.5-2.5 mM OA; concentration step: 0.5 mM OA; applied potential value of 1.65V/SCE.

Chapter 12

ELECTROANALYTICAL APPLICATION OF CARBON-BASED COMPOSITE ELECTRODES

Selected organic molecules were chosen to represent target pollutants in aqueous solutions, *i.e.*, thiourea for sulfur-based impurities, glucose for carbohydrates, and urea for nitrogen-based impurities. In addition, phenol derivates, e.g., 4-chlorophenol (4-CP), 2,4-dinitrophenol (2,4-DNP), and 2,4-dichlorophenol (2,4-DCP) as priority pollutant set by EU and by USEPA [67, 68] and oxalic acid as a result of the partial degradation of 4-chlorophenol were selected as well. Also, the simultaneous detection of organics from different classes has been reported. The advanced composite electrodes have been employed for the detection of these analytes under different conditions (see Table 8).

INDIVIDUAL DETECTION OF ORGANICS AT COMPOSITE-BASED ELECTRODES

Individual Detection of Phenol Derivates

Even if the phenols are oxidisable at relatively low anodic potentials in the aqueous media, making electrochemistry a viable method for their degradation and detection, the phenol electrooxidation leads to electrode fouling by electropolymerization that leads to a low rate of oxidation, low permeability, and strong adhesion on the electrode surface. For this reason, the electrode renewal was performed by mechanical polishing and electrochemical pre-treatment by repetitive cycling scanning or pre-anodization at a fixed potential value. Cyclic voltammetry (CV), linear-scan voltammetry (LSV), chronoamperometry (CA), differential-pulsed voltammetry (DPV), and multiple-pulsed voltammetry (MPA) were used for the 4-CP determination on advanced composite electrodes. The use of DPV has shown to be very sensitive for detection of organic molecules [114,115].

Table 8. Individual detection of organics at advanced composite electrodes

Composite electrode type	Supporting electrolyte	Analyte	References
EGE	0.1 M Na_2SO_4	4-CP	[39, 43, 46, 50]
	0.1 M Na_2SO_4	2,4-DNP	-
	0.1 M Na_2SO_4	OA	[39]
	-	TU	[49]
EGPS	0.1 M Na_2SO_4	4-CP	[39, 43, 62]
	0.1 M Na_2SO_4	2,4-DNP	-
	0.1 M Na_2SO_4	OA	[39, 47]
	-	TU	[49]
AgZEGE	0.1 M Na_2SO_4/0.1 M NaOH	4-CP	[43]
	0.1 M NaOH	2,4-DCP	-
	0.1 M NaOH	urea	[51]
CuZEGE	0.1 M Na_2SO_4/0.1 M NaOH	4-CP	[44]
	0.1 M NaOH	2,4-DCP	-
	0.1 M NaOH	glucose	-

The electroanalytical performances of 4-CP detection in aqueous solution using EGE and EGPS electrodes are sumarised in Table 9.

Table 9. Electroanalytical performances for 4-CP detection in 0.1 M Na_2SO_4 supporting electrolyte using EGE and EGPS electrodes

Electrode type	Used technique	Concentration range (mM)	Sensitivity ($mA \cdot mM^{-1} \cdot cm^{-2}$)	Correlation coefficient (R^2)	LOD (mM)	RSD* (%)
EGE	CV	0.1-0.5	0.39	0.980	0.01	2.8
	LSV	0.1-0.5	0.38	0.982	0.01	2.7
	DPV	0.1-0.5	0.40	0.986	0.01	2.9
	CA	0.5-2.5	0.05	0.992	0.1	2.9
EGPS	CV	0.1-0.5	0.31	0.980	0.01	2.9
	LSV	0.1-0.5	0.31	0.982	0.01	2.8
	DPV	0.1-0.5	0.35	0.986	0.01	2.7
	CA	0.5-2.5	0.04	0.992	0.1	3.0

*RSD-relative standard deviation was determined for three replicates

Among presently used electroanalytical techniques, the simpliest is CA, but because of electrode fouling occurring during the detection, multiple-pulsed amperometry (MPA) has proved to be extremely sensitive for the detection of organic molecules because of the advantage of *in-situ* cleaning and re-activating the electrode surface during the electroanalytical detection [116]. In Figures 31 and 32 are shown examples of the results of CA and respective, MPA applying.

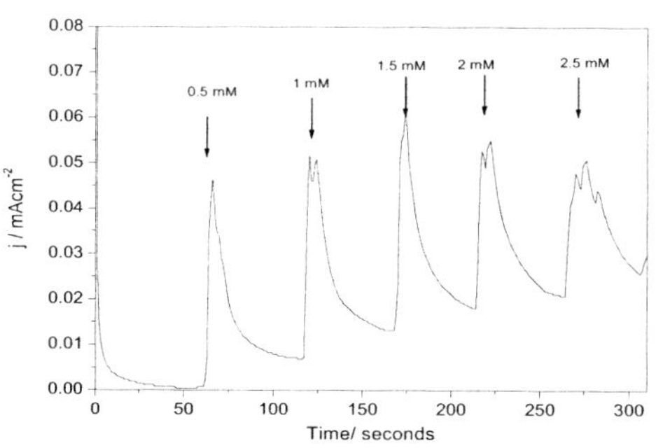

Figure 31. Chronoamperogram (CA) of the EGE electrode in 0.1 M Na_2SO_4 supporting electrolyte and in the presence of different 4-CP concentrations recorded at E= +1.1V vs. SCE.

A positive cleaning potential ($E_{oxidation}$) that removes the oxidizable contaminant on the electrode surface, is applied for the time period $t_{oxidation}$, and there were notable improvements, compared to the condition of constant potential. Introduction of MPA led to a good improvement, in terms of higher sensitivity. Thus, applying MPA improved significantly the detection performance, owing to the electrode cleaning during the detection experiments and a better sensitivity (0.24 $mAmM^{-1}cm^{-2}$), compared with the sensitivity obtained by using CA (0.05 $mAmM^{-1}cm^{-2}$).

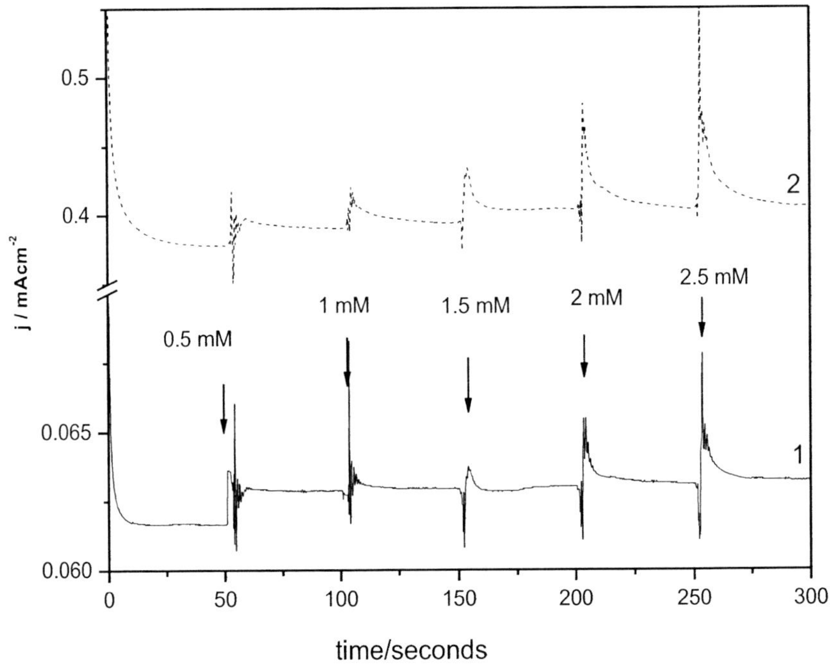

Figure 32. Multiple pulsed amperogram (MPA) of EGE electrode in 0.1 M Na_2SO_4 supporting electrolyte and in the presence of different 4-CP concentrations; $E_{detection}$ was +1.1 V(curve 1) and $E_{oxidation}$ was +1.4 V vs. SCE (curve 2); $t_{detection}$ was 0.05 seconds and $t_{oxidation}$ was 0.15 seconds.

The use of zeolite-modified electrodes also improves the sensitivity of the oxidation processes of the compound. The chemical modification of the electrodes is a field of growing interest in analytical chemistry. In electrochemistry, it has been demonstrated that zeolite-modified electrodes

(ZMEs) possess distinct advantages over conventional electrodes in many application areas, including electrocatalysis and electrochemical sensors [40].

AgZEGE and CuZEGE were employed comparatively with an EGE electrode for 4-CP determination under the conditions of 0.1 M Na_2SO_4 and 0.1 M NaOH supporting electrolyte (see Table 10). Under the conditions of 0.1 M Na_2SO_4 supporting electrolyte the AgZEGE electrode exhibited the oxidative/reductive peaks corresponding to redox peaks of the Ag/Ag(I) couple. For this electrode, the background current is higher than for the EGE electrode. In the presence of 4-CP, the oxidation-process overpotential of 4-CP for AgZEGE electrode was 0.05 V smaller than for the EGE electrode, and the anodic peak of Ag oxidation decreased in the presence of 4-CP, possibly due to an adsorption effect. Under these conditions, no enhancement of detection method of 4-CP was reached. In 0.1 M NaOH supporting electrolyte, some differences in relation with EGE occurred. Figures 33 a-c show examples of differential-pulsed voltammograms on the three types of electrodes in a 0.1 M NaOH supporting electrolyte. It can be seen that the presence of 4-CP influenced the occurrence of the anodic peaks corresponding to the electroformation of soluble $[Ag(OH)_2]^-$ complex species and the electroformation of Ag_2O (the first and second peaks), and to the formation of AgO (third peak). It has to be mentioned that only for DPV applying at the AgZEGE electrode in 0.1 M NaOH to detect 4-CP, the electrode surface needed to be renewed by mechanical polishing and three cycling scannings between –0.5 V to 0.8 V vs. SCE in supporting electrolyte to reach the reproducible results. Under these conditions, the electrode sensitivity decreased, possibly due to the presence of 4-CP hampering the formation of oxidized Ag species, and the lack of cathodic potential, which not allowed the electrogeneration of the reduced form of mobile silver (Figure 33 a). Using CuZEGE electrode, two peaks at about 0.35 and 0.44 V vs. SCE were observed, corresponding to the electrochemical oxidation of reduced Cu^+ [55] and respective, 4-CP oxidation (Figure 33 b). In comparison with CV, the DPV peak of 4-CP is much more pronounced. However, the electrode sensitivities and the correlation coefficients determined at the potential values corresponding to the redox process of CuO were slightly lower compared to the results obtained at the potential value corresponding to the 4-CP oxidation on graphite, but the better lowest limit of detection was reached.

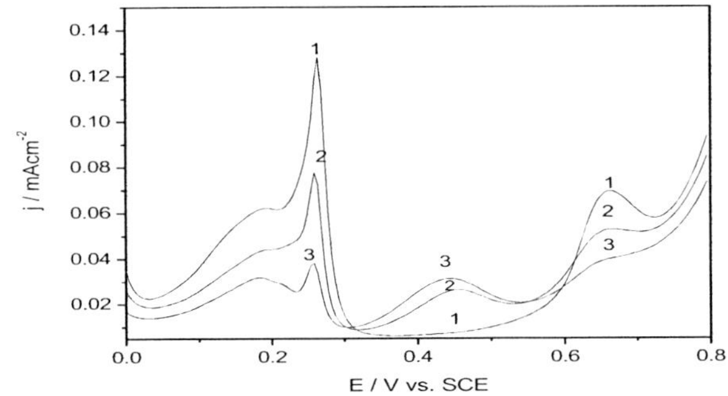

a)

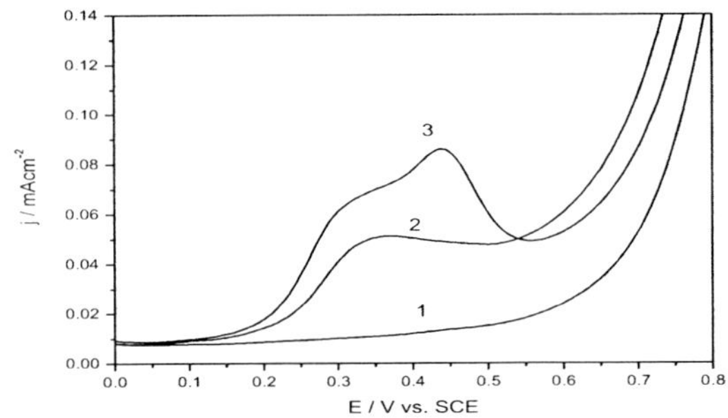

b)

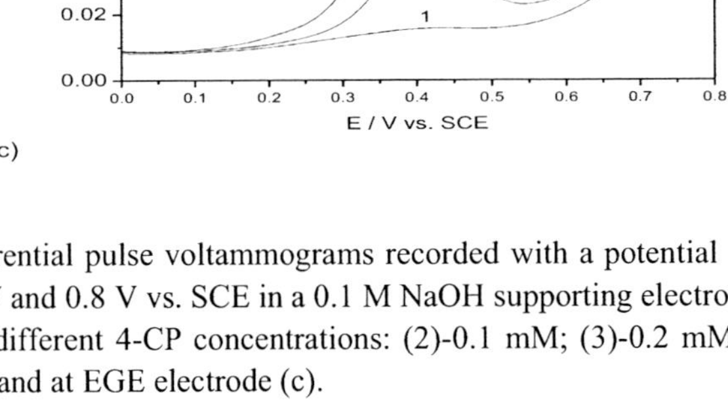

c)

Figure 33. Differential pulse voltammograms recorded with a potential scan rate 0.05 Vs^{-1} between 0 V and 0.8 V vs. SCE in a 0.1 M NaOH supporting electrolyte (1) and in the presence of different 4-CP concentrations: (2)-0.1 mM; (3)-0.2 mM; at AgZEGE (a), CuZEGE(b) and at EGE electrode (c).

Table 10. The electroanalytical parameters of the amperometric detection of 4-CP in alkaline supporting electrolytes at EGE, AgZEGE and CuZEGE electrodes using different techniques

Oxidation potential (V vs. SCE)	Electrode type	Technique used	Electrode sensitivity ($mA \cdot mM^{-1} \cdot cm^{-2}$)	Correlation coefficient (R^2)	LOD (mM)	RSD* (%)
0.5	EGE	CV	0.24	0.982	0.01	2.8
0.5	EGE	DPV	0.24	0.986	0.01	2.7
0.5	AgZEGE	CV	0.22	0.983	0.01	2.4
0.5	AgZEGE	DPV	0.22	0.989	0.01	2.2
0.5	CuZEGE	CV	0.30	0.990	0.013	3.0
0.35	CuZEGE	DPV	0.12	0.930	0.001	1.2
0.44	CuZEGE	DPV	0.18	0.980	0.006	2.5

* for three replicates

Under these working conditions, no significant improvement of electroanalytical performances of ZMEs compared with EGE electrodes subjected to electrode sensitivity for 4-CP detection was achieved. Though, the copper presence exhibited its catalytic effect by the lower overpotential of 4-CP oxidation at 0.35 V/SCE, and better LOD. In order to enhance the electronalaytical response for 4-CP detection, the next informations show the results of applying a preconcentration-voltammetric detection scheme.

For using the ZMEs in a preconcetration-voltammetric detection scheme the accumulation time is very important, because it could influence the degree of adsorption on the electrode surface. An important role that may be played by the zeolite is its availability to concentrate species within its porous structure. This would also increase the amount of species available for reaction. 4-CP is a polarisable specie and should be adsorbed or even interact with a more hydrophilic zeolite, although a more hydrophobic zeolite exhibits a higher adsorption capacity. An additional explanation has also been proposed for the oxidation of hydroquinone due to the ability of a zeolite to orient molecules in a way to increase the rate of the redox reaction [117-120]. Also, for this situation of zeolite doped with Ag or Cu, the catalyst type influences the rate of the redox reaction. To determine the enhancement factor as the ratio of the peak current density observed for the ZME to that for the corresponding EGE electrode [120], the effect of accumulation time on anodic peak current densities was investigated by CV for AgZEGE in 0.1 M Na_2SO_4 supporting electrolyte and for CuZEGE in 0.1 M NaOH supporting electrolyte, and the results are presented in Figure 34. In the case of the EGE electrode no enhancement of the anodic peak current density was achieved, while for the ZMEs for an increasing accumulation time the amount of 4-CP at the electrode surface increased leading to the enhancement of the anodic peak current density. The enhancement factor of about 16 for AgZEGE and about 6.5 for CuZEGE revealed an effective concentration effect of 4-CP on ZMEs. Also, it is well-known that the kinetics of the accumulation process depends on the nature of the zeolite-modified electrode. The accumulation time of 12 and respectively, 15 minutes required to reach the equilibrium reveals the presence of a diffusion effect of solution molecules in the bulk of the electrodes, and ascribed to the electrode (surface) porosity [43]. The renewal of the electrode used in 4-CP detection was achieved by chemical polishing, without necessity of the chemical regeneration, though. The difference between the enhancements factors obtained for both types of ZMEs (Table 11) should be related to the electrode porosity, proved also by the longer accumulation time to reach the equilibrium and alkaline medium that influenced the electrode surface by the oxides within different zeolite's sites.

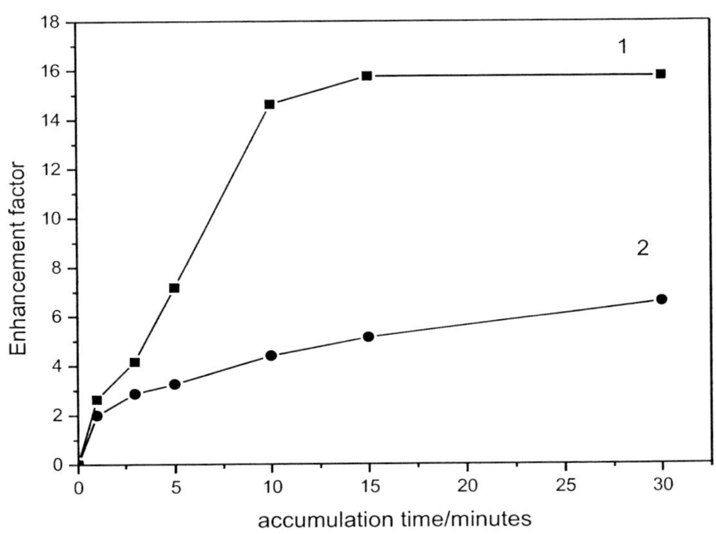

Figure 34. The enhancement factor obtained by a preconcentration-voltammetric detection scheme using CV versus the accumulation time for the AgZEGE electrode in 0.1 M Na_2SO_4 (curve 1) and for CuZEGE electrode in 0.1 M NaOH supporting electrolyte (curve 2).

Table 11. Electroanalytical performance of ZMEs for the detection of 4-CP using preconcentration-voltammetric detection scheme

Oxidation potential (V vs. SCE)	Electrode type	Supporting electrolyte	Electrode sensitivity ($mA \cdot mM^{-1} \cdot cm^{-2}$)	Correlation coefficient (R^2)	LOD (mM)	RSD* (%)
0.7	AgZEGE	0.1 M Na_2SO_4	0.79	0.992	0.001	1.8
0.5	CuZEGE	0.1 M NaOH	0.56	0.991	0.01	1.1

* for three replicates.

2,4-Dinitrophenol (2,4-DNP) is also a "priority phenol", a representant of the di-substituted phenol derivate class, with electrochemical hehaviour different versus monosubstituted phenol derivates. The electrochemical behaviour of 2,4-DNP at EGE, EGPS and CFEG electrodes are studied by CV. Figures 35 a-c show the cyclic voltammograms for 0.2 and 1 mM 2,4-DNP in 0.1 M Na_2SO_4 supporting electrolyte at EGE, EGPS and CFEG electrodes. The shapes of cyclic voltammograms are different for the studied electrodes. Even the anodic current increasing with 2,4-DNP concentration at high overpotential very near to starting oxygen evolution, only for EGPS an limiting current appeared at the potential value of +1.23 V vs. Ag/AgCl. For each tested electrodes the anodic current decreased different with cycling number (the results are not shown here).

In addition, it can be noticed that increasing 2,4-DNP concentration led to the cathodic current increasing. For each 2,4-DNP concentration the cathodic current size is different in relation with the type of the composite electrode, as the type of insulating matrix (epoxy or polystyrene) and the content of conductive carbon filler. Under this cathodic potential range, the formation of aminophenol derivates can occur, and converted further into aminophenol-based polymer. This polymer is less stabil than mono-substituted nitrophenol based polymer, being further oxidized and even mineralized [111].

Based on CV results obatined in the presence of 2,4-DNP, the multiple-pulsed amperometry (MPA) technique was used under different conditions in comparison with CA to improve the electrochemical detection of 2,4-DNP at carbon-based composite electrodes. In Figure 36 is shown an example of chronoamperogram registered for EGE electrode at the detection potential of +1.25 V vs. Ag/AgCl, and an electrode fouling effect was noticed at 2,4-DNP concentrations higher than 0.4 mM.

a)

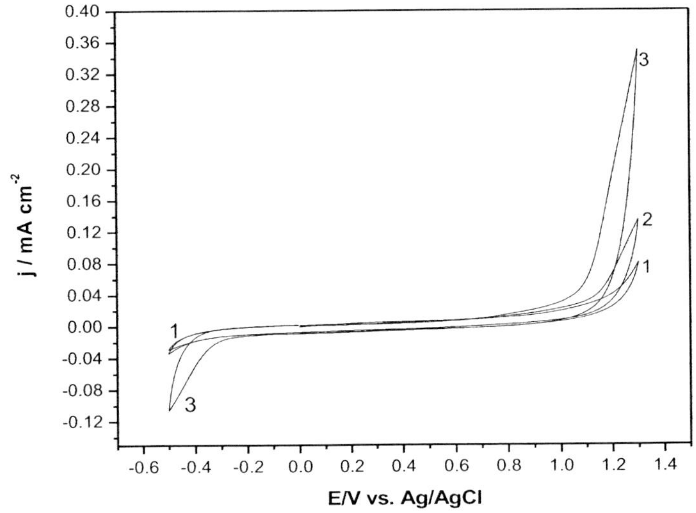
0.40
0.36
0.32
0.28
0.24
0.20
0.16
0.12
0.08
0.04
0.00
-0.04
-0.08
-0.12
j / mA cm^{-2}
-0.6
-0.4
-0.2
0.0
0.2
0.4
0.6
0.8
1.0
1.2
1.4
E/V vs. Ag/AgCl
1
2
3

b)

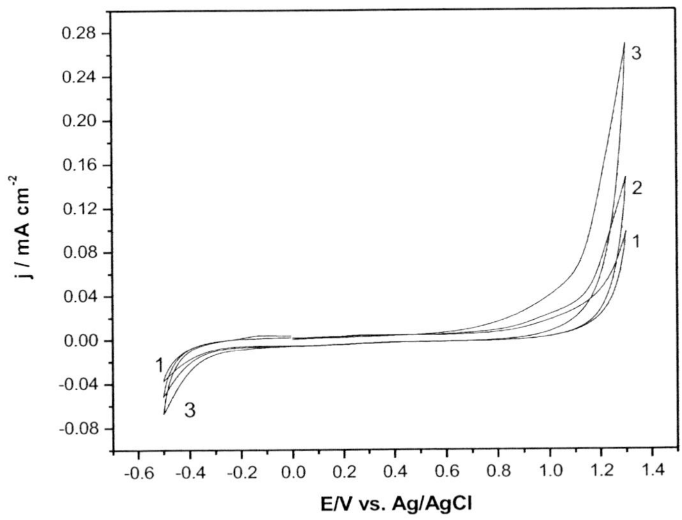
0.28
0.24
0.20
0.16
0.12
0.08
0.04
0.00
-0.04
-0.08
j / mA cm^{-2}
-0.6
-0.4
-0.2
0.0
0.2
0.4
0.6
0.8
1.0
1.2
1.4
E/V vs. Ag/AgCl
1
2
3

c)

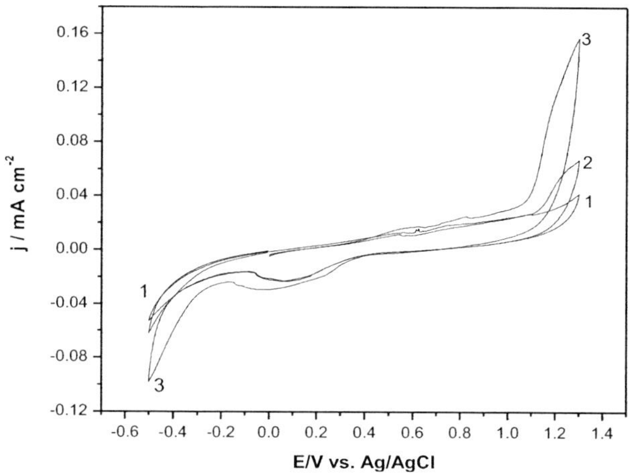

Figure 35. Cyclic voltammograms in 0.1 M Na_2SO_4 supporting electrolyte (curve 1) and in the presence of 0.2 mM 2,4-DNP (curve 2) and 1 mM 2,4-DNP (curve 3) at: a-EGE, b-CFEGE, c-EGPS electrodes; starting potential: 0 V vs. Ag/AgCl; potential range: -0.5 V→ +1.3 V → -0.5V vs. Ag/AgCl; scan rate: 0.05 Vs-1.

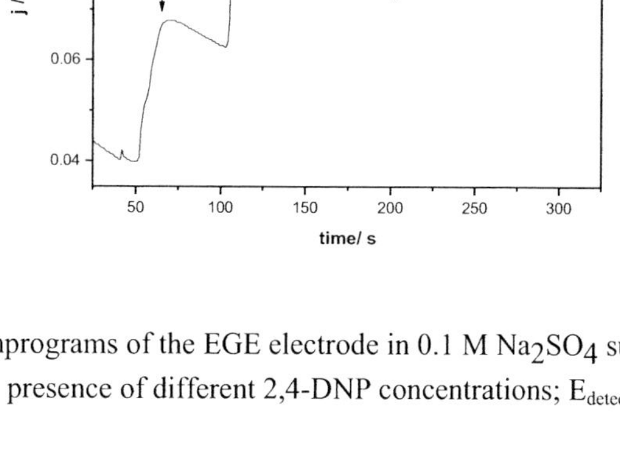

Figure 36. Chronoamprograms of the EGE electrode in 0.1 M Na_2SO_4 supporting electrolyte and in the presence of different 2,4-DNP concentrations; $E_{detection}$ was +1.25 V vs. Ag/AgCl.

Thus, an optimum protocol for eliminating the fouling effect (Figure 36), should be investigated by using MPA technique. In Table 12 are gathered the variants of MPA applying.

Table 12. MPA applying for the electrochemical detection of 2,4-DNP at carbon-based composite electrodes

MPA type	Number of levels/ type of level	Potential /V vs. Ag/AgCl	Time /s
MPA_I	I – detection	+1.25	0.05
	II - cleaning	+1.5	0.15
MPA_{II}	I - detection	+1.25	0.05
	II - cleaning	-0.5	0.15
MPA_{III}	I – detection	+1.25	0.05
	II – cleaning	+1.5	0.15
	III - conditioning	-0.5	0.15

The results obtained for EGE electrode expressed as amperograms by MPA applying at different conditions are presented in Figures 37 and 38 as examples for two and three levels of potential.

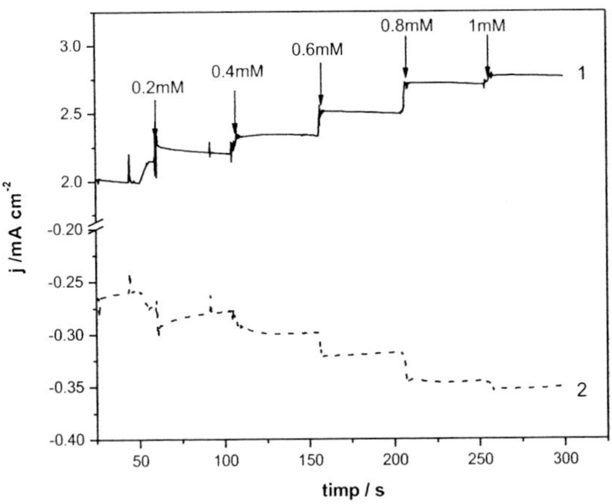

Figure 37. Multiple-pulsed amperogram (MPA) of the EGE electrode in 0.1 M Na_2SO_4 supporting electrolyte and in the presence of different 2,4-DNP concentrations; $E_{detection}$ was +1.25 V vs. Ag/AgCl (curve 1) and $E_{cleaning}$ was -0.5 V vs. Ag/AgCl(curve 2); $t_{detection}$ was 0.05 s and $t_{cleaning}$ was 0.15 s.

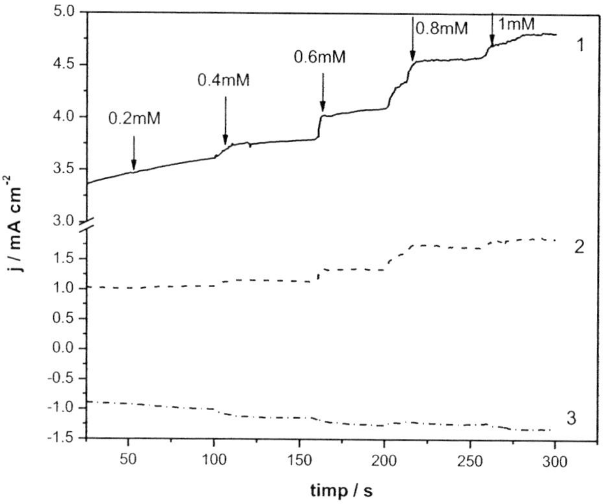

Figure 38. Multiple- pulsed amperograms (MPAs) of EGE electrode in 0.1 M supporting electrolyte and in the presence of different 2,4-DNP concentrations; Edetection was +1.25 V vs. Ag/AgCl (curve 1), Ecleaning was+1.5 V (curve 2), Econditioning was -0.5 V vs. Ag/AgCl (curve 3); tdetetion was 0.05 s; tcleaning was 0.15 s, tcond was 0.15 s.

Under the concentration range between 0.2 and 1 mM 2,4-DNP applying CA at fixed potential value of + 1.25 V vs. Ag/AgCl, no linear dependence between current density and 2,4-DNP concentration was achieved. MPA applying under the conditions of three levels corresponding to different potential values allowed to reach a linear dependence, and a high sensitivity for 2,4-DNP detection. The potential values are chosen based on cyclic voltammetry behaviour of 2,4-DNP at each electrode versus the oxidation potential and the reduction potential. Under these conditions, a very high enhancement of the sensitivity was reached (see Table 13).

The electrochemical determination of another di-substituted phenol derivate, 2,4-DCP was tested at AgZEGE and CuZEGE electrodes in 0.1 M NaOH supporting electrolyte. In comparison with 4-CP detection at AgZEGE electrode (the oxidation potential at +0.5 V vs. SCE) [43], the cyclic voltammograms for different concentrations of 2,4-DCP (Figure 39) show two oxidation potential at +0.4 and respective, +0.6 V vs. SCE.

Table 13. The electroanalytical performances carbon based electrodes for the electrochemical detection of 2,4-DNP versus applied conditions

Electrode type	Technique		Sensitivity/ mA $cm^{-2}mM^{-1}$	
	CA Parameters	MPA Parameters	CA	MPA
EGE	E= +1.25 V	$E_{oxidation}$=+1.25 V; $t_{oxidation}$=0.05 s $E_{cleaning}$=+1.5 V; $t_{cleaning}$=0.15s		*
		$E_{oxidation}$=+1.25 V; $t_{oxidation}$=0.05 s $E_{cleaning}$= -0.5 V; $t_{cleaning}$=0.15s	*	0.7224
		$E_{oxidation}$=+1.25 V; $t_{oxidation}$=0.05 s $E_{cleaning}$=+1.5 V; $t_{cleaning}$=0.15s $E_{conditioning}$=-0.5 V; $t_{conditioning}$=0.15 s		1.6172
CFEGE	E= +1.25 V	$E_{oxidation}$=+1.25 V; $t_{oxidation}$=0.05 s $E_{cleaning}$=+1.5 V; $t_{cleaning}$=0.15s		*
		$E_{oxidation}$=+1.25 V; $t_{oxidation}$=0.05 s $E_{cleaning}$= -0.5 V; $t_{cleaning}$=0.15s	*	0.1358
		$E_{oxidation}$=+1.25 V; $t_{oxidation}$=0.05 s $E_{cleaning}$=+1.5 V; $t_{cleaning}$=0.15s $E_{conditioning}$=-0.5 V; $t_{conditioning}$=0.15 s		1.5701
EGPS	E= +1.25 V	$E_{oxidation}$=+1.25 V; $t_{oxidation}$=0.05 s $E_{cleaning}$=+1.5 V; $t_{cleaning}$=0.15s		0.0271
		E_{ox}=+1.25 V; t_{ox}=0.05 s $E_{cleaning}$= -0.5 V; $t_{cleaning}$=0.15s	*	0.1149
		$E_{oxidation}$=+1.25 V; $t_{oxidation}$=0.05 s $E_{cleaning}$=+1.5 V; $t_{cleaning}$=0.15s $E_{conditioning}$=-0.5 V; $t_{conditioning}$=0.15 s		0.1719

* no linear dependence of current density versus 2,4-DNP concentration.

For both oxidation potentials, the linear dependences of the peak currents versus 2,4-DCP concentrations were achieved. At the oxidation potential of +0.6 V vs. SCE the electroanalytical parameters expressed as the sensitivity (0.2 $mAcm^{-2}mM^{-1}$) and the correlation coefficient (0.995) were better than recorded at +0.4 V vs. SCE (the sensitivity of 0.09 $mAcm^{-2}mM^{-1}$ and the correlation coefficient of 0.986). Also, the electroanalytical parameters obtained for di-substituted and mono-substituted chlorophenols in relation with the oxidation potential, sensitivity and concentration range were similar (Table 14).

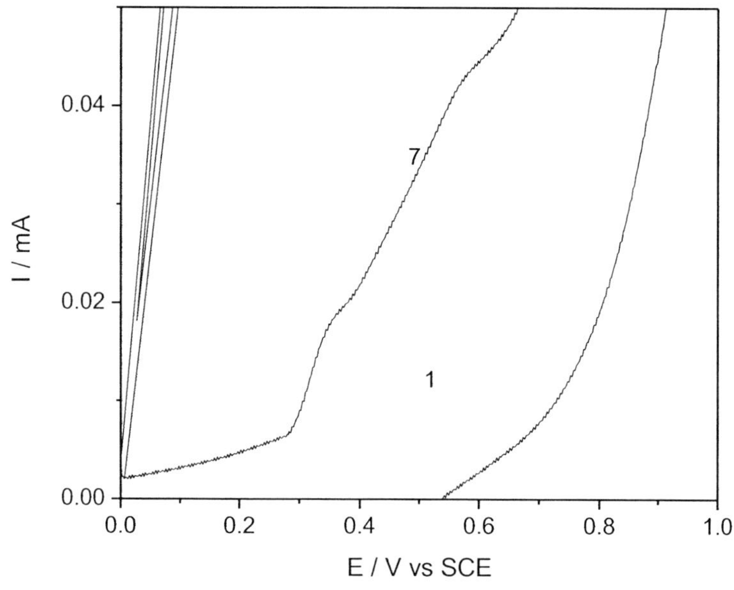

Figure 39. Detail of cyclic voltammograms of AgZEGE electrode in 0.1 M NaOH supporting electrolyte (curve 1) and in the presence of various concentrations of 2,4-DCP: 0.1 mM to 0.6 mM (curves 2-7), concentration step: 0.1 mM 2,4-DCP; potential scan rate: 0.05 Vs^{-1}.

Another example of 2,4-DCP detection in alkaline medium is shown for use of CuZEGE electrode (Figure 40). No significant difference versus the use of AgZEGE electrode was reached.

Table 14. Comparative results for the electroanalytical parameters for the electrochemical detection of 4-CP and 2,4-DCP at ZME electrodes using CV in alkaline media

ZME type	Oxidation potential	Phenol derivate	Concentration range (mM)	Sensitivity ($mAmM^{-1}cm^{-2}$)	Correlation coefficient (R^2)
AgZEGE	+0.5	4-CP	0.1-0.5	0.22	0.983
AgZEGE	+0.6	2.4-DCP	0.1-0.6	0.20	0.995
CuZEGE	+0.5	4-CP	0.1-0.5	0.27	0.990
CuZEGE	+0.6	2,4-DCP	0.1-0.6	0.27	0.977

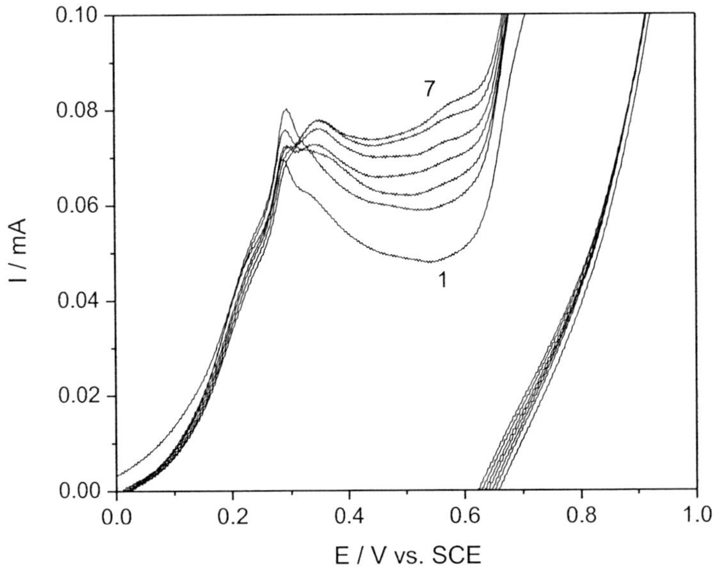

Figure 40. Detail of cyclic voltammograms of CuZEGE electrode in 0.1 M NaOH supporting electrolyte (curve 1) and in the presence of various concentrations of 2,4-DCP: 0.1 mM to 0.6 mM (curves 2-7), concentration step: 0.1 mM 2,4-DCP; potential scan rate: 0.05 Vs-1.

Electrochemical Detection of Oxalic Acid (OA)

The advanced EGE and EGPS composite electrodes allowed to detect OA in aqueous solutions. A comparative example of electrode behaviour in the presence of OA for both electrodes is shown detailed in Figure 41 and no significant difference in relation with the useful signal for 0.1 mM OA concentration was noticed. The electroanalytical parameters depended on the electrochemical used (see Table 16) and not on the electrode type characterized by the different insulating matix. Some examples of CVs at an EGPS and DPVs at an EGE electrode are presented in Figures 42 and 43.

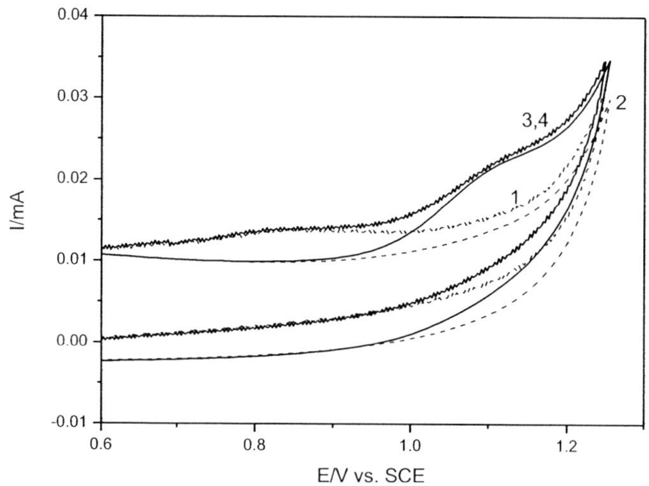

Figure 41. CVs of the EGPS and EGE electrodes in a 0.1 M Na_2SO_4 supporting electrolyte (curves 1, 2) and in the presence of 0.1 mM OA (curves 3, 4); potential scan rate of 0.05 Vs^{-1}.

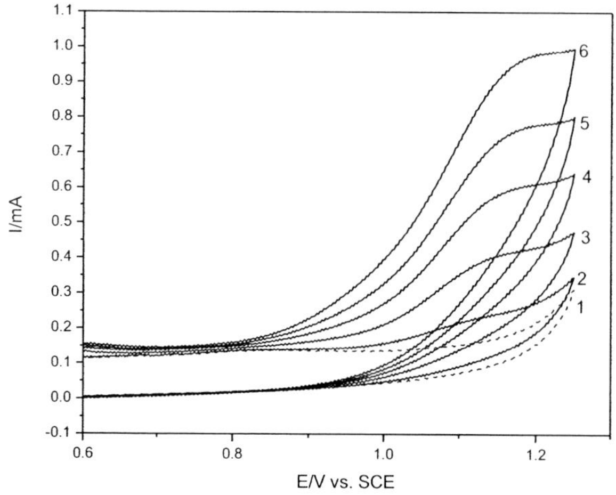

Figure 42. CVs of the EGPS electrode in 0.1 M Na_2SO_4 supporting electrolyte (1) and in the presence of different OA concentrations: 2-0.1 mM OA, 3-0.3 mM OA, 4-0.5 mM OA, 5-0.7 mM OA, 6-0.9 mM OA; potential scan rate of 0.05 Vs^{-1}.

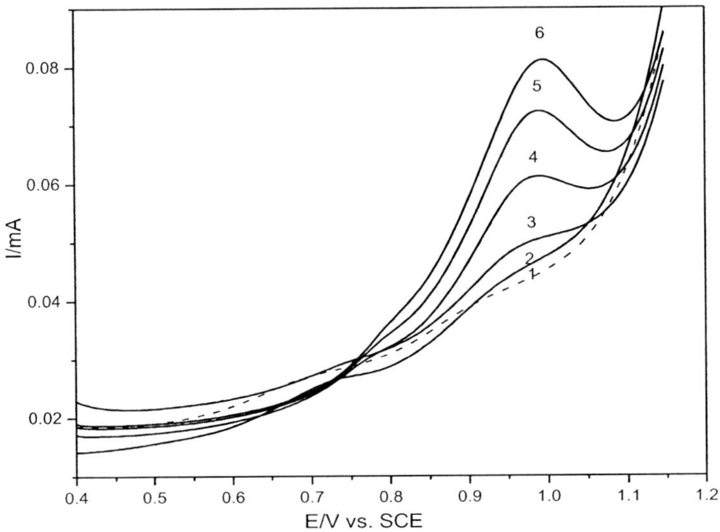

Figure 43. DPVs of the EGE electrode in 0.1 M Na_2SO_4 supporting electrolyte (1) and in the presence of different OA concentrations: 2-6: 0.1-0.5 mM OA.

Table 16. The electroanalytical parameters of voltammetric/amperometric detection of OA using EGE and EGPS electrodes in 0.1 M Na_2SO_4 supporting electrolyte

Electrode type	Concentration range (mM)	Technique used	Electrode sensitivity ($mA \cdot mM^{-1} \cdot cm^{-2}$)	Correlation coefficient (R^2)	LOD (mM)	RSD* (%)
EGE	0.1-1	CV	1.10	0.998	0.01	2.6
	0.1-0.7	DPV	0.98	0.990	0.01	3.1
	0.5-3	CA	0.20	0.989	0.1	3.2
EGPS	0.1-1	CV	1.11	0.998	0.05	2.5
	0.1-1	LSV	0.96	0.992	0.08	2.8
	0.5-3	CA	0.18	0.988	0.1	3.2

*for five replicates.

The comparative detection results for OA determination using the two types of electrodes revealed a similar electroanalytical performance of the electrodes. The precision of the electrochemical determination of OA was evaluated using conventional $KMnO_4$ titration method and the accuracy of this method is very good [47].

Electrochemical Detection of Urea

The determination of urea is essential for municipal wastewaters control, because of very fast and high expansion of industrial and farming activity. In municipal wastewaters the organic nitrogen specie is represented by urea and about 80% the nitrogen in fresh urine is bound in urea [121]. Also, its determination is very important for a number of applications, e.g., clinical diagnosis, food processing, agricultural processes and environmental protection [122-126]. For the analysis of urea in aqueous samples several analytical procedures have been developed. Most of these investigations have been based on measuring the changes of ammonia enzymatically released during the hydrolysis of urea. There are several analytical procedures based on optical methods [127, 128] and electrochemical techniques. In general, the determination of urea by electrochemical techniques means the use of enzyme-modified electrodes by potentiometric and amperometric methods [122, 124, 125]. Though, a non-enzimatic amperometric detection of urea in aqueous solution at poly(Ni-cyclam) film-modified glassy carbon electrode was reported by Ferrer [123].

From the advanced electrode materials tested, AgZEGE exhibited useful features for the electrochemical determination of urea [51].

A series of resultant CVs obtained directly in the concentration range 0.1 mM – 2.6 mM for urea standard solutions, corresponding to curves 1-6, is shown in Figure 44. The anodic current peaks corresponding to AgO formation increased progressively linearly with urea concentration. Also, the two cathodic current peaks recorded at about 0 and 0.2 V vs. SCE increased with urea concentrations, but the current within this potential range was not considered for the electrochemical detection due to the fact it was not stable in the absence of urea (the results are not shown here). A very unexpected and unusual oxidation peak occurred immediately after the reduction of AgO to Ag_2O on backward scanning in cathodic sense, which increased also linearly with urea concentration. This behaviour could be the result of continuously remove of the surface layers of Ag_2O in the presence of ammonium as a product of catalyzed hydrolysis of urea with formation of soluble complex of diammine silver ($Ag(NH_3)_2^+$), thus allowing the oxidation to proceed as the electrode is not oxidized in depth [129]. This explanation was supported by the linear voltammetric results obtained in the presence of different urea concentrations, with a non-linear dependence of the current peak height vs. urea concentration. Also, the comparative results in the presence of urea and ammonium ions showed the similare behaviour [51].

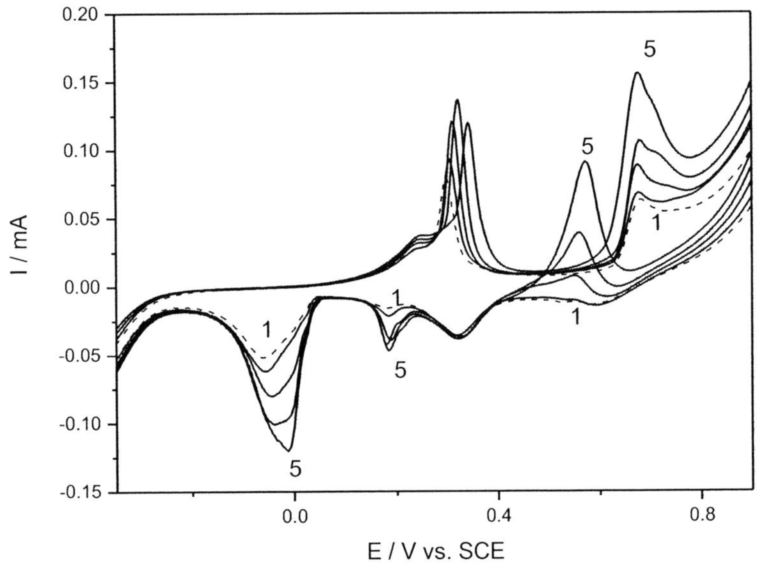

Figure 44. Cyclic voltammograms of the AgZEGE composite electrode in a 0.1 M NaOH supporting electrolyte (1) and in the presence of different urea concentrations: 2-0.1 mM; 3-0.2 mM; 4-0.3 mM; 5-0.4 mM; 6-0.6 mM; 7-0.8 mM; 8-1 mM; 9-1.2 mM; 10-1.4 mM; 11-1.8 mM; 12-2.2 mM; 13-2.6 mM; potential scan rate 0.05 Vs-1 ; potential range:-0.5 to 1.25 V.

This modified electrode exhibited an electrocatalytic effect via the redox couple of AgO and Ag_2O towards urea oxidation allowing its determination in aqueous solution. The electrochemical determination of urea on the AgZEGE electrode was obatained by all techniques used, i.e., CV, LSV, DPV, SWV, CA. The electronalytical sensitivity for the determination of urea ranged from 0.153 to 0.295 $mA{\cdot}mM^{-1}{\cdot}cm^{-1}$ depending on the technique used (Table 17). The analytical singnals were recorded after preliminary conditioning at the potential value of +1.25V/SCE for 30 seconds in the presence of urea. No analytical signals occurred without preconditioning, an aspect that strengthens the idea that the signal corresponds to a soluble complex of diammine silver.

From CA data, when the concentration of urea was higher than 1.4 mM, a response platform as a "catalytic saturation" was observed, showing a characteristic in formal accordance with the Michaelis-Menten catalytic kinetic mechanism (the results are not shown here). The kinetic parameters were obtained from the electrochemical version of the Lineweaver-Burk equation [130]. The apparent Michelis-Menten type constant, K'_M , and I_{max} were determined by analysis of the slope and intercept of the plot of the reciprocals of the steady-state current recorded at 50 seconds versus urea concentration. The K'_M value of the AgZEGE electrode was 1.6 mM and j_{max} =0.255 mA $\cdot cm^{-1}$ Also, the sensitivity calculated as j_{max}/K'_M ratio was 0.153 $mA \cdot mM^{-1} \cdot cm^{-1}$.

Table 17. The electroanalytical performance of AgZEGE electrode for the detection of urea for the concentration range of 0.2 to 1.4 mM [38]

Peak potential	Technique used	Sensitivity ($mA \cdot mM^{-1} \cdot cm^{-1}$)	Correlation coefficient (R^2)
+0.68	CV	0.168	0.994
+0.55*	CV	0.224	0.993
+0.41	SWV	0.295	0.990
+0.38	DPV	0.224	0.991
+0.55	CA	0.153	0.999

* recorded on the reverse scanning of CV.

A moderate enhancement of electroanalytical sensitivity (0.331 $mA \cdot mM^{-1} \cdot cm^{-1}$) for the determination of urea at the AgZEGE composite electrode was reached by applying a chemical preconcentration step prior to voltammetric/amperometric quantification. The enhancement factor of about 3 reached by increasing accumulation time up to 15 minutes revealed an effective concentration effect of AgZEGE electrode on urea.

Electrochemical Detection of Glucose

In the last years, development of glucose electrochemical detection method using copper based electrodes has attracted a growing attention [6,

131-140], because the copper in alkaline media can act as an electrocatalyst for the oxidation of organic species, which are mainly responsible for COD.

The CuZEGE electrode was tested for detection glucose in a 0.1 M NaOH supporting electrolyte. Based on CV behaviour (Figure 45) the global oxidation process of glucose on CuZEGE could be complex process mediated by the Cu (I), Cu (II) species in the potential range from 0 to +1.25V/SCE. Based on obtained results, especially by second oxidation peak appearance at more positive potentials, and by analogy with the data from literature [140], the glucose oxidation mechanism which involves also the redox couple Cu(III)/Cu(II) should not be regarded as a simple mediated electrochemical process. The electrocatalytic activity of copper-based electrodes in alkaline solutions probably involves adsorbed hydroxyl radicals $(HO^{\bullet})_{ads.}$. These highly reactive species, generated through the anodic discharge of hydroxide ion by Cu(II) oxide sites, represents an intermediate product of the oxygen evolution [140].

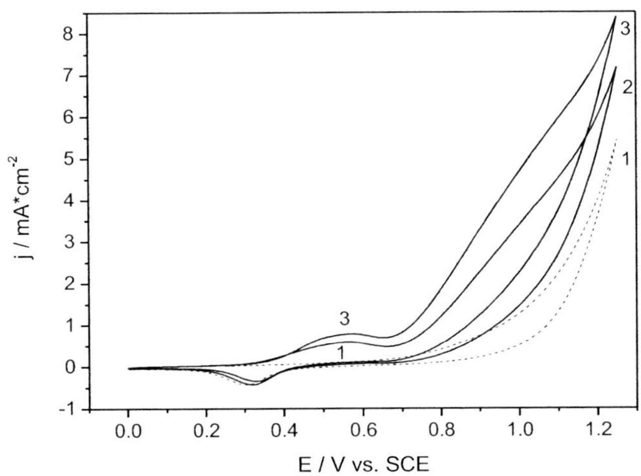

Figure 45. Cyclic voltammograms of CuZEGE electrode recorded at a potential scan rate of 0.05 Vs-1 from 0 to +1.25 V vs. SCE in 0.1 M NaOH supporting electrolyte (curve 1), in the presence of 5 mM glucose (curve2), and in he presence of 8 mM glucose (curve 3).

Thus, in the electrocatalytic process the source of O atoms transferred to the oxidation products are probably $(HO^{\bullet})_{ads}$ species. Even the oxidation process of glucose starts at lower potential value (about +0.5 V vs. SCE), a lower sensitivity for the electrochemical detection was obtained, with fast electrode fouling. Based on these results, for the electrochemical determination of glucose, the second potential value of about +1 V vs. SCE was tacking into consideration.

This electrode allowed determining electrochemically glucose using CV, DPV, and CA. In Figure 46 a series of DPVs in 0.1 M NaOH supporting electrolyte and in the presence of different concentration of glucose are presented. By applying a pre-concentration scheme for the CuZEGE electrode in the presence of glucose an enhancement of an useful signal was not achieved, the electroanalytical performance being similar with the ones obtained without preconcentration step. The enhancement factor of about 1.14 practically demonstrates the absence of an effective concentration effect of the CuZEGE electrode on glucose. Therefore, using a preconcentration-detection scheme is not justified for this analyte. The electroanalytical parameters obtained for the electrochemical determination of glucose using the above-mentioned techniques are gathered in Table 18.

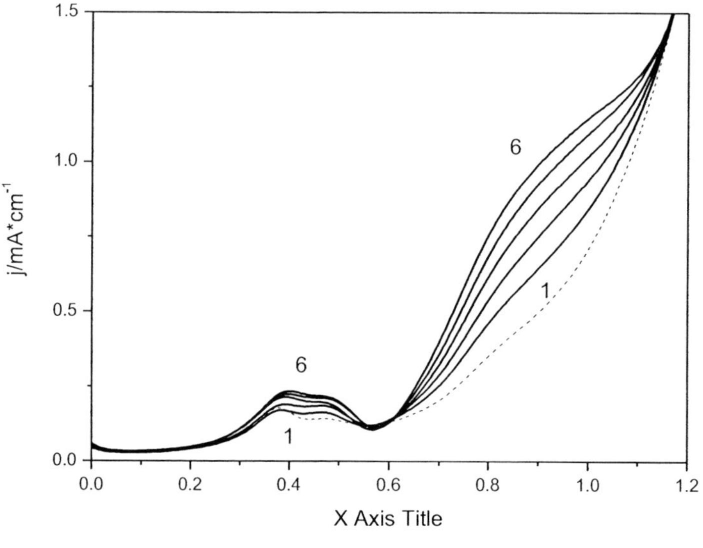

Figure 46. Differential pulse voltammograms of the CuZEGE electrode recorded at a potential scan rate of 0.05 Vs^{-1} from 0 to +1.25 V vs. SCE in 0.1 M NaOH supporting electrolyte (1) and in the presence of different glucose concentrations: 2-6:1-5 mM

Table 18. Electroanalytical performance of the CuZEGE electrode for the detection of glucose in 0.1 M NaOH solution

Peak potential (V/SCE)	Technique used	Concentration range (mM)	Sensitivity ($mA \cdot mM^{-1} \cdot cm^{-2}$)	Correlation coefficient (R^2)	LOD (mM)	RSD * (%)
+1.00	CV	1-9	0.42	0.997	0.37	2.6
+0.90	DPV	1-5	0.10	0.993	0.5	2.7
+0.90	CA	1-7	0.12	0.990	0.06	2.1
+0.90	Preconcentration/ DPV	1-5	0.12	0.996	0.4	2.7

* for three replicates.

Chapter 13

Electrochemical Simultaneous Detection of Organics At Composite Electrodes

Simultaneous Detection of Oxalic Acid (OA) and 4-Chlorophenol (4-CP)

In general, chlorophenols and saturated carboxylic acids are refractory compounds to conventional chemical and biological oxidation [108]. The oxygenated compounds, such as saturated carboxylic acid, are the intermediates, or the end products of the conventional and advanced oxidation process of chlorophenols contained in waste waters [117-120]. Under these conditions, the simultaneous determination of chlorophenol and saturated carboxylic acid in water is required especially to elucidate the mechanism of oxidation process. The presence of saturated carboxylic acid, e.g., oxalic acid (OA) in water after the application of the oxidation process of 4-CP shows that no complete mineralization occurred. The determination of saturated carboxylic acid as intermediary or end product for the 4-CP oxidation by using HPLC [121] and electrochemical methods [25, 37, 48] was reported.

EGE electrode was tested to detect OA in the presence of 4-CP. Based on CVs obtained in the presence of each individual organic compound (discussed in the above section), the oxidation peak potentials of OA and 4-CP of 1.1 and respective, 0.8 V/SCE. Although, this electrode exhibited a better sensitivity for 4-CP determination, the presence of OA was also detected.

Figure 47 shows an example of differential pulse voltammograms for single organic and the mixture of OA and 4-CP at EGE electrode. Using CV,

CA, and DPV, OA was detected in the presence of 4-CP with no requirement of complicated procedures.

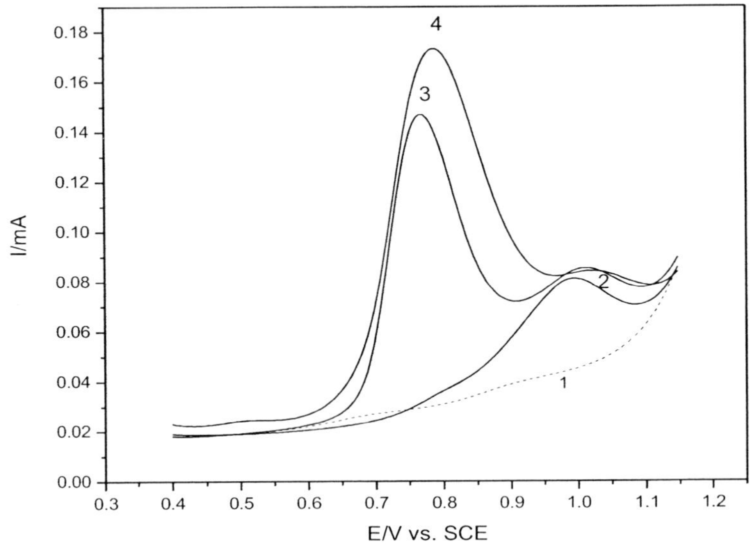

Figure 47. Differential- pulsed voltammograms of the EGE electrode in a 0.1 M Na_2SO_4 supporting electrolyte (curve 1), in the presence of 0.5 mM OA(curve 2), in the presence of 0.5 mM 4-CP and 0.5 mM OA (curve 3), in the presence of 0.6 mM 4-CP and 0.5 mM OA (curve 4).

Also, by using chronoamperometry, the detection of OA from the mixture of 4-CP and OA was achieved, even if the loss of the electrode activity occurred due to the electrode fouling at the potential value corresponding to the 4-CP oxidation (0.8 V/SCE). The useful net current signals recorded at 1.1 V /SCE (after 50 seconds) are linearly dependent on the OA concentration in the explored concentration range between 0.1 mM and 1 mM mixture (Figure 48). For the mixture with every 4-CP and OA concentrations beyond 0.5 mM, the electrode activity is decreased.

The comparative sensitivity for electrochemical determination of OA in the presence/absence of 4-CP by using CV, DPV, and CA is gathered in Table 19.

A practical application and a recovery test was also performed by analyzing the tap water samples spiked with 25 mg·dm^{-3} 4-CP and 14.2 mg·dm^{-3} Na_2SO_4 before and after the electrochemical oxidation at the EGPS electrode with a geometric surface area of 81 mm^2 under the galvanostatic conditions (j= 0.04 $mA \cdot cm^{-2}$, t=2h) using the chronopotentiometric method [48].

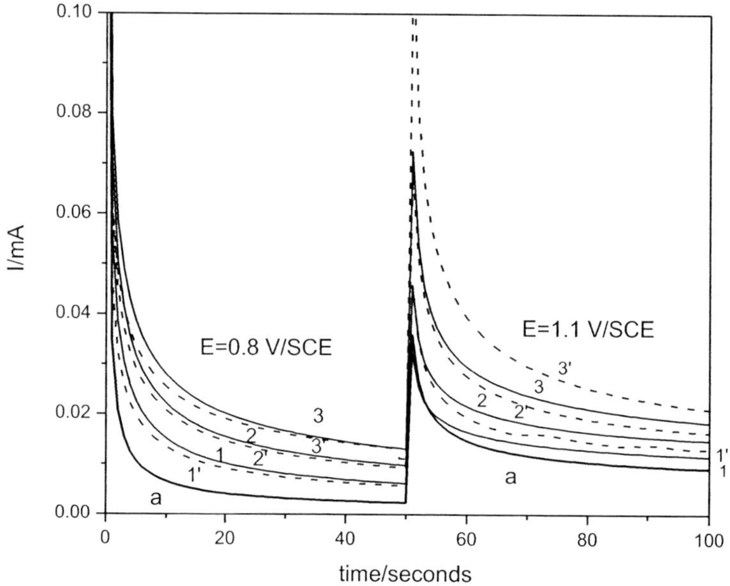

Figure 48. Chronoamperograms of the EGE electrode in 0.1 M Na_2SO_4 supporting electrolyte (curve a) and the presence of different concentrations of 4-CP(-), curves 1-3: 0.1 mM to 0.3 mM and the mixture with different concentrations of 4-CP and OA(--), curves 1'-3': 0.1 mM to 0.3 mM.

Table 19. Comparative sensitivities for OA determination in the absence/presence of 4-CP

Concentration range (mM)	Technique used	Electrode sensitivity ($mAmM^{-1}cm^{-2}$)	
		no presence of 4-CP	presence of 4-CP
0.1-1	CV	1.10	1.27
0.1-0.7	DPV	0.98	1.16
0.5-3	CA	0.20	0.18

The sample analysis was carried out using the standard addition method by DPV technique, and the results are summarized in Table 20. By this practical application, the duality of the advanced composite electrode materials was demonstrated, by using for the electrochemical degradation of organic pollutant and for the electrochemical determination of these and its oxidation intermediar/product.

Table 20. Determination of 4-CP, OA in tap water sample containing 14.2 gdm^{-3} Na_2SO_4 before and after electrochemical oxidation at EGPS electrode using DPV technique and EGE electrode

Sample	Species	Detected* (mg dm^{-3})	Added (mg dm^{-3})	Found* (mg dm^{-3})	RSD (%)	Recovery (%)
Initial	4-CP	25.6	25	48.6	4.2	96
	OA	-	18	18.6	3.5	103.3
After electrochemical oxidation	4-CP	21.5	25	45.3	4.6	97.4
	OA	9.8	18	26.8	4.9	96.4

*Mean value ±standard deviation (n=3)

Finally, the results obtained by this method were compared with those obtained by means of a conventional $K_2Cr_2O_7$ titration method as Chemical Oxygen Demand (COD) parameter. Using the titration method to evaluate the precision of this method, it can be concluded that the results obtained by the two methods are concordant and the accuracy of the proposed voltammetric method is good.

Simultaneous Detection of Glucose and 2,4-Dichlorophenol (2,4-DCP)

Taking into account the simultaneous presence of different pollutants classes in water, ZME electrodes were tesetd to detect simultaneously the glucose as representant of the carbohydrate class in the presence of phenol derivate. Figures 49 and 50 show two examples of simultaneous detection of glucose in the presence of 2,4-DCP at AgZEGE electrode using CV and at CuZEGE electrode using DPV.

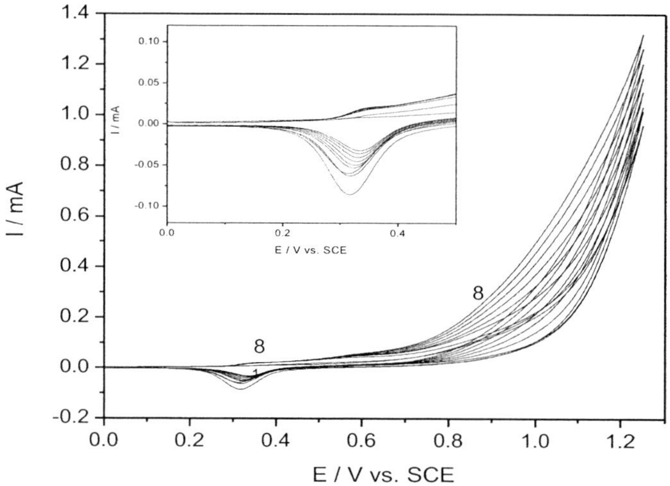

Figure 49. Cyclic voltammograms of the AgZEGE electrode in 0.1 M NaOH supporting electrolyte (curve 1), in the presence of 0.3 mM 2,4-DCP(curve 2), and in the presence of 0.6 mM DCP in the mixture with different glucose concentrations: 3-8: 1-6 mM glucose. Details of the limited potential range characteristics to 2,4-DCP detection.

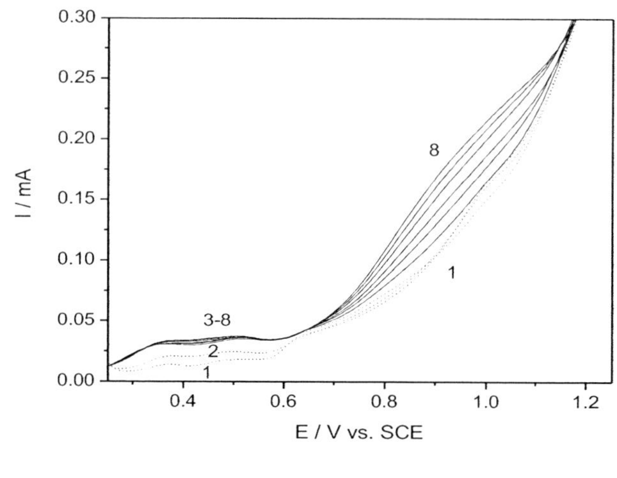

Figure 50. Differential-pulsed voltammograms of the CuZEGE electrode in a 0.1 M NaOH supporting electrolyte (curve 1), in the presence of 0.1 mM 2,4-DCP(curve 2), in the presence of 0.6 mM 2,4-DCP and different glucose concentrations: 3-8: 1-6 mM glucose.

SIMULTANEOUS DETECTION OF GLUCOSE AND UREA

Another example of simultaneous electrochemical detection of pollutants, which belongs to different class of organics responsible for organic load (glucose and urea) is shown in Figure 51. The simultaneous detection of glucose and urea can be performed at AgZEGE electrode, based on the above-presented results regarding urea detection. Using DPV, the sensitivity for the electrochemical determination of glucose in the absence /presence of urea was almost similar (0.1 versus 0.07 $mAmM^{-1}cm^{-2}$). For the simultaneous electrochemical detection of glucose and urea, a prior conditioning stage was neccesary. This stage consisted of a preanodization at the potential value of +1.5 V/SCE for 60 seconds in the presence of urea was established during the electrochemical determination of urea. This could exploit further for the selective detection of glucose in the presence of urea. Applying the same conditions without the preanodization stage, the urea was not detected (the results are not shown here).

a)

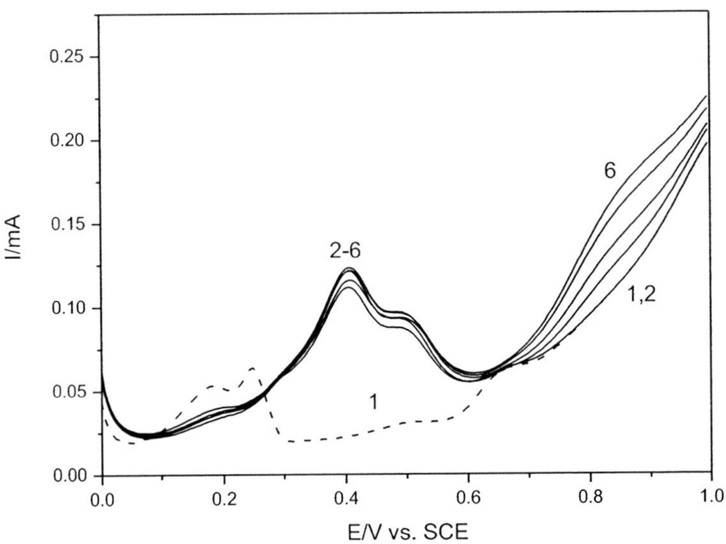

b)

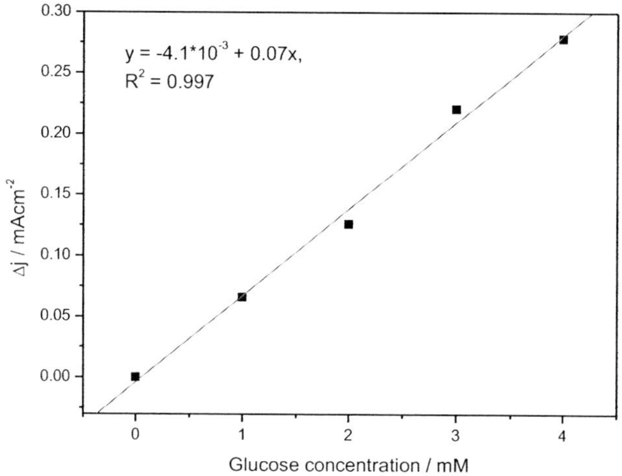

Figure 51. a) Differential pulse voltammograms of the AgZEGE electrode in a 0.1 M NaOH supporting electrolyte (curve 1), in the presence of 2 mM urea (curve 2), and in the presence of different glucose concentrations: 3-6: 1-4 mM glucose; scan rate of 0.05 Vs^{-1}. b) Calibration curve recorded at +0.85 V/SCE.

Chapter 14

CONCLUSION

From the different examples above-described in relation with advanced electrode material behaviour and application for the electrochemical determination of some selected organic compounds, it is evidenced the possibility of using advanced-carbon based composite electrodes as electrochemical sensors. These electrodes exhibited useful features for the electrochemical determination of selected pollutants with respect to its mechanical resistance, very low cost, simple preparation and easy renewal of the active electrode surface.

The use of film casting technique is an easy and useful technique for obtaining stable electrodes consisting of composite material.

The comparative electrochemical characterization of boron-doped diamond, C-composite and zeolite modified electrodes showed the widest potential window and the lowest background current for BDD electrode.

The different electrical and electrochemical behaviours of the composites electrodes are based on the morphological differences based on the different natures of the insulating matrices and the carbon flakes orientation and loading.

Under certain conditions, the expanded graphite-epoxy (EGE), expanded graphite-polystyrene (EGPS) composite electrodes proved the electrochemical behaviour corresponding to microelectrode array, and exhibited the ability to deliver the current responses for an electroactive specie in the absence of any supporting electrolyte.

Improvement of the electron transfer rate at carbon surface was achieved by the deliberate alteration of carbon surface modified zeolite. The Romanian natural zeolite with major content of clinoptilolite can be used as support for a catalyst, e.g., copper or silver, in order to improve the performance of electroanalytical detection. For certain pollutants, the enhancement of electroanalysis performance was achieved by the presence of zeolite, by using Ag–modified zeolite-expanded graphite-epoxy (AgZEGE) and Cu-modified zeolite-expanded graphite-epoxy (CuZEGE) composite electrode as zeolite modified electrodes (ZMEs).

During the detection of phenol derivates, all tested electrodes exhibited sorption and surface process, which led to the electrode fouling. The avoidance of this effect was reached applying multiple-pulsed amperometry technique at certain potential levels, which assured an electrochemical in-situ polishing.

The use of boron-doped diamond electrode to detect certain compounds, e.g., oxalic acid, as-freshed electrode had to be used. No detection of oxalic acid can be achieved at oxidized BDD electrode.

The electrocatalytic properties of ZMEs were proved towards oxidation and determination of urea and glucose, which can not be determined in the absence of catalyst.

Due to the fast response and stability of these electrodes, they can be proposed also as detectors in flow system.

References

[1] Zhao, H; Jiang, D; Zhang, S; Catterall, K; John, R. Development of a Direct Photoelectrochemical Method for Determination of Chemical Oxygen Demand. *Analytical Chemistry* 2004, 76, 155-160.

[2] Yu, H; Wang, H; Quan, X; Chen, S; Zhang, Y. Amperometric Determination of Chemical Oxygen Demand Using Boron-Doped Diamond (BDD) Sensor. *Electrochemistry Communications* 2007, 9, 2280-2285.

[3] Li, J; Li, L; Zheng, L; Xian, Y; Ai, S; Jin, L. Amperometric Determination of Chemical Oxygen Demand with Flow Injection Analysis Using F-PbO_2 Modified Electrode. *Analytica Chimica Acta* 2005, 548, 199-204.

[4] Zhang, J; Zhou, B; Zheng, Q; Li, J; Bai, J; Liu, Y; Cai, W. Photoelectrocatalytic COD Determination Method Using Highly Ordered TiO_2 Nanotube Array. *Water Research* 2009, 43, 1986-1992.

[5] Manea, F; Radovan, C; Vlaicu, I. Electrochemical Assessment of Amperomtric Oxidability Level in Water Pollution Monitoring. *Chemical Bulletin of "Politehnica" University of Timisoara, Romania* 2005, 50 (64), 5-8.

[6] Lee, KH; Ishikawa, T; McNiven, SJ; Nomura, Y; Hiratsuka, A; Sasaki, S; Arikawa, Y; Karube, I. Evaluation of Chemical Oxygen Demand (COD) Based on Coulometric Determination of Electrochemical Oxygen Demand (EOD) Using a Surface Oxidized Copper Electrode. *Analytica Chimica Acta* 1999, 398, 161-172.

[7] Westbroek, P; Temmerman, E. In Line Measurement of Chemical Oxygen Demand by Means of Multipulse Amperometry at a Rotating Pt-Ring-Pt/PbO_2 Disc Electrode. *Analytica Chimica Acta* 2001, 437(1), 95-105.

[8] Tu, Y; Lin, Y; Yantasee, W; Ren, Z. Carbon Nanotube Based Nanoelectrode Arrays: Fabrication, Evaluattion, and Application in Voltammetric Analysis. *Electroanalysis* 2005, 17, 79-84.

[9] Sarada, BV; Terashima, C; Ivandini, TA; Rao, TN; Fujishima, A. Electroanalytical Applications of Highly Boron-Doped Diamond Electrode. In: Fujishima A, Einaga Y, Rao TN, Tryk DA. *Diamond Electrochemistry*. Tokyo and Amsterdam. BKC INC and Elsevier BV; 2005; 261-286.

[10] Spataru, N; Tryk, DA; Fujishima, A. Anodic Voltammtery at Conductive Diamond Electrodes and Its Analytical Applications. In: Fujishima A, Einaga Y, Rao TN, Tryk DA. *Diamond Electrochemistry*. Tokyo and Amsterdam. BKC INC and Elsevier BV; 2005; 287-308.

[11] Pleskov, YuV. Electrochemistry of Diamond: A Review. *Russian Journal of Electrochemistry* 2002, 38(12), 1411-1429.

[12] Kraft, A. Doped Diamond: A Compact Review on a New, Versatile Electrode Material. *International Journal of Electrochemical Science* 2007, 2, 355-385.

[13] Martin, HB; Argoitia, A; Angus, JC; Landau, U. Voltammetry Studies of Single-Crystal and Polycrystalline Diamond Electrodes. *Journal of the Electrochemical Society* 1999, 146 (8), 2959-2964.

[14] Ivandini, TA; Rao, TN; Fujishima, A; Einaga, Y. Electrochemical Oxidation of Oxalic Acid at Highly Boron-Doped Diamond Electrode. *Analytical Chemistry* 2006, 78, 3467-3471.

[15] Salazar-Banda, GR; Andrade, LS; Nascente, PAP; Pizani, PS; Rocha, RC; Avaca, LA. On the Changing Electrochemical Behaviour of Boron-Doped Diamond With Time After Cathodic Pre-treatments. *Electrochimica Acta* 2006, 51(22), 4612-4619.

[16] Granger, MC; Xu, J; Strojek, JW; Swain, GM. Polycrystalline Diamond Electrodes: Basic Properties and Applications as Amperometric Detectors in Flow Injection Analysis and Liquid Chromatography. *Analytica Chimica Acta* 1999, 397, 145-161.

[17] Wang, J. In Analytical Electrochemistry; VCH Publishers: New York, 2000; Chapter 4, p.100-139.

[18] Rassaei, L; Sillanpaa, M; Bonne, MJ; Marken, F. Carbon Nanofiber-Polystyrene Composite Electrodes for Electroanalytical Processes. *Electroanalysis* 2007, 19(14), 1461-1466.

[19] Zhu, L; Tian, C; Zhai, J; Yang, R. Sol-Gel Derived Carbon Nanotubes Ceramic Composite Electrodes for Electrochemical Sensing. *Sensors and Actuators B* 2007, 125, 254-261.

[20] Pumera, M; Merkoci, A; Alegret, S. Carbon Nanotube-epoxy Composites for Electrochemical Sensing. *Sensors and Actuators B* 2006, 113, 617-622.

[21] Ballarin, B; Cordero-Rando, MM; Blanco, E; Hidalgo-Hidalgo De Cisneros, JL; Seeber, R; Tonelli, D. New Rigid Conducting Composites for Electrochemical Sensors. *Collection of Czechoslovak Chemical Communications* 2003, 68, 1420-1436.

[22] Ramirez-Garcia, S; Alegret, S; Cespedes, F; Forster, RJ. Carbon Composite Electrodes: Surface and Electrochemical Properties. *Analyst* 2002, 127, 1512-1519.

[23] Gabay, T; Ben-David, M; Kalifa, I; Sorkin, R; Abrahams, ZR; Ben-Jacob, E; Hanein, Y. Electro-Chemical and Biological Properties of Carbon Nanotubes Based Multi-Electrode Arrays. *Nanotechnology* 2007, 18, 035201-035206.

[24] Manso, J; Mena, ML; Yanez-Sedeno, P; Pingarron, J. Electrochemical Biosensors Based on Colloidal Gold-nanotube Composite Electrodes. *Journal of Electroanalytical Chemistry* 2007, 603 (1), 1-7.

[25] Sun, D; Zhu, L; Zhu, G. Glassy Carbon Ceramic Composite Electrodes. *Analytica Chimica Acta* 2006, 564, 243-247.

[26] Mendes, RK; Cervini, P; Cavalheiro, ETG. The Use of a Graphite-Castor Oil Polyurethane Composite Electrode for the Determination of Hydroquinone in Photographic Developers. *Talanta* 2006, 68, 708-712.

[27] O'Hare, D; Macpherson, JV; Wilson, A. On the Microelectrode Behaviour of Graphite-Epoxy Composite Electrodes. *Electrochemistry Communications* 2002, 4, 245-250.

[28] Marken, F; Gerrard, ML; Mellor, IM; Mortimer, RJ; Madden, CE; Fletcher, S; Holt, K; Foord, JS; Dahm, F. Voltammetry at Carbon Nanofiber Electrodes. *Electrochemistry Communications* 2001, 3, 177-180.

[29] Barsan, MM; Pinto, EM; Florescu, M; Brett, CMA. Development and Characterization of a New Conducting Carbon. *Analytica Chimica Acta* 2009, 635, 71-78.

[30] Simm, AO; Banks, CE; Ward-Jones, S; Davies, TJ; Lawrence, NS; Jones, TGJ; Jiang, L; Compton, RG. Boron-Doped Diamond Microdisc Arrays: Electrochemical Characterisation and Their Use as a Substrate for the Production of Microelectrode Arrays of Diverse Metals (Ag, Au, Cu) via Electrodeposition. *Analyst* 2005, 130, 1303-1311.

[31] Stulik, K; Amatore, C; Holub, K; Marecek, V; Kutner, W. Microelectrodes. Definitions, Characterization, and Applications. *Pure Applied Chemistry* 2000, 72(8), 1483-1492

[32] Davies, TJ; Compton, RG. The Cyclic and Linear Sweep Voltammetry of Regular and Random Arrays of Microdisc Electrodes: Theory. *Journal of Electroanalytical Chem*istry 2005, 585, 63-82.

[33] Cięciwa, A; Wuthrich, R; Comninellis, C. Electrochemical Characterization of Mechanically Implanted Boron-Doped Diamond Electrodes. *Electrochemistry Communications* 2006, 8, 375-382.

[34] Soh, KL; Kang, WP; Davidson, JL; Basu, S; Wong, YM; Cliffel, DE; Bonds, AB; Swain, GM. Diamond-Derived Microelectrodes Array for Electrochemical Analysis. *Diamond Related Materials* 2004, 13, 2009-2015.

[35] Feeney, R; Kounaves SP. Microfabricated Ultramicroelectrode Arrays: Developments, Advances, and Applications in Environmental Analysis. *Electroanalysis* 2000, 12(9), 677-684.

[36] Fletcher, S; Horne, MD. Random Assemblies of Microelectrodes (RAM electrodes) for Electrochemical Studies. *Electrochemistry Communications* 1999, 1, 502-512.

[37] Lawrence, NS; Pagels, M; Meredith, A; Jones, TGJ; Hall, CE; Pickles CSJ; Godfried, HP; Banks, CE; Compton RG; Jiang, L. Electroanalytical Applications of Boron-Doped Diamond Microelectrode Arrays. *Talanta* 2006, 69, 829-834.

[38] Lee, HJ; Beriet, C; Ferrigno, R; Girault, H. Cyclic Voltammetry at Regular Microdisc Electrode Array, *Journal of Electroanalytical Chemistry* 2001, 502, 138-145.

[39] Manea, F; Radovan, C; Pop, A; Corb, I; Burtica, G; Malchev, P; Picken, S; Schoonman, J. Carbon Composite Electrodes Applied for Electrochemical Sensors. In: Baraton, M.I. *Sensors for Environment, Health and Security*. Dordrecht: Springer; 2009; 179

[40] Walcarius, A. Implication of Zeolite Chemistry in Electrochemical Science and Applications of Zeolite-modified Electrodes. In: Auerbach, SM; Carrado, KA; Dutta, PK. *Handbook of zeolite science and technology*. New York: Marcel Dekker; 2003; 721.

[41] Granda Valdés, M; Pérez-Cordoves, AI; Díaz-García ME. Zeolites and Zeolite Based Materials in Analytical Chemistry. *TrAC Trends in Analytical Chemistry* 2006, 25 (1), 24-30.

[42] Walcarius, A. Zeolite-Modified Paraffin-Impregnated Graphite Electrode. *Journal of Solid State Electrochemistry* 2006, 10, 469-478.

[43] Pop, A; Manea, F; Radovan, C; Malchev, P; Bebeselea, A; Proca, C; Burtica, G; Picken, S; Schoonman, J. Amperometric Detection of 4-Chlorophenol on Two Types of Expanded Graphite Based Composite Electrodes. *Electroanalysis* 2008, 20(22), 2460-2466.

[44] Pop, A; Manea, F; Malchev, P; Proca, C; Burtica, G; Picken, S; Schoonman, J. Voltammetric Detection of 4-Chlorophenol on Copper-Zeolite-Expanded Graphite-Epoxy Composite Electrodes. *Journal of Optoelectronics and Advanced Materials-Symposia* 2009, 1(1), 96-100.

[45] Chen, G; Weng, W; Wu, D; Wu C. PMMA/Graphite Nanosheets Composite and Its Conducting Properties. *European Polymer Journal* 2003, 39(12), 2329-2335.

[46] Pop, A; Manea, F; Radovan, C; Corb, I; Burtica, G; Malchev, P; Picken, S; Schoonman, J. Determination of 4-Chlorophenol Using Two Types of Graphite-Based Composite Electrodes. *Revue Romaine de Chemie* 2008, 53(8), 623-628.

[47] Manea, F; Radovan, C; Corb, I; Pop, A; Burtica, G; Malchev, P; Picken, S; Schoonman, J. Electrochemical Oxidation and Determination of Oxalic Acid at an Expanded-Polystyrene-Composite Electrode. *Sensors* 2007, 7, 615-627.

[48] Orha, C; Manea, F; Pop, A; Burtica, G; Fazekas-Todea I. Obtaining and characterization of zeolitic materials with antibacterial properties. *Revista de Chimie* 2008, 59, 173-177.

[49] Corb, I; Manea, F; Radovan, C; Pop, A; Burtica, G; Plamen, M; Picken, S; Schoonman, J. Carbon-Based Composite Electrodes: Preparation, Characterization and Application in Electroanalysis. *Sensors* 2007, 7, 2626-2635.

[50] Manea, F; Radovan C; Corb, I; Pop, A; Burtica, G; Malchev, P; Picken, S; Schoonman, J. Simultaneous Determination of 4-Chlorophenol and Oxalic Acid Using an Expanded Graphite-Epoxy Composite Electrode. *Electroanalysis* 2008, 20(5), 1719-1724.

[51] Manea, F; Pop, A; Radovan, C; Malchev, P; Bebeselea, A; Burtica, G; Picken, S; Schoonman, J. Voltammetric Detection of Urea at Ag-Modified Zeolite-Expanded Graphite-Epoxy Composite Electrode. *Sensors* 2008, 8, 5806-5819.

[52] Mentus, SV; Bošković, I; Pješčić, JM; Grudić, V; Bogdanov, Ž. Tailoring the Morphology and Electrocatalytic Properties of Electrochemically Formed Ag/TiO2 Composite Deposits on Ttanium Surfaces. *Journal of Serbian Chemical Society* 2007, 72 (12), 1403-1418.

[53] Jovič, BM; Jovič, VD. Electrochemical Formation and Characterization of Ag_2O. *Journal of Serbian Chemical Society* 2004, 69 (2), 153-166.

[54] Manea, F; Radovan, C; Schoonman, J. Amperometric Determination of Thiourea in Alkaline Media on a Copper Oxide-Copper Electrode. *Journal of Applied Electrochemistry* 2006, 36, 1075-1081.

[55] Li, J; Calzaferri, G. Copper-Zeolite-Modified Electrodes: An Intrazeolite Ion Transport Mechanism. *Journal of Electroanalytical Chemistry* 1994, 377, 163-175.

[56] Laviron, E. General Expression of the Linear Potential Sweep Voltammogram Diffusionless Electrochemical Systems. *Journal of Electroanalytical Chemistry* 1979, 101, 19-28.

[57] Filipe, OMS; Brett, CMA. Characterization of Carbon Film Electrodes for Electroanalysis by Electrochemical Impedance. *Electroanalysis* 2004, 16 (12), 994-1001.

[58] Watanabe, T; Ivandini, TA; Makide, Y; Fujishima, A; Einaga Y. Selective Detection Method Derived from a Controlled Diffusion Process at Metal-Modified Diamond Electrodes. *Analytical Chemistry* 2006, 78, 7857-7860.

[59] Codognoto, L; Machado, SAS; Avaca, LA. Square Wave Voltammetry on Boron-Doped Diamond Electrodes for Analytical Determinations. *Diamond Related Materials* 2002, 11, 1670-1675.

[60] Merriman, JC; Anthony, DHJ; Kraft, JA; Wilkinson, RJ. Rainy River Water Quality in the Vicinity of Bleached Kraft Mills. *Chemosphere* 1991, 23, 1605-1615.

[61] Kang, MG.; Jung, HS.; Kim, KL. Effect of Chloride Ions on 4-Chlorophenol Photodegradation in the Absence and Presence of Titanium Silicalite-2. *Journal of Photochemistry and Photobiology A* 2000, 136, 117-123.

[62] Manea, F; Radovan, C; Burtica, G; Bebeselea, A; Pop, A; Corb, I; Schoonman, J. Degradation of 4-Chlorophenol from Wastewaters by Electrooxidation Using Graphite Based Composite Electrodes. *Environmental Science and Technology* 2007, 1, 281-287.

[63] Kristiansen, NK; Froshaug, M; Aune, KT; Becher, G. Identification in Chlorinated Seawater and Drinking Water Produced Offshore Using Open-Loop Stripping Technique. *Environmental Science and Technology* 1994, 28, 1669-1673.

[64] DeWalle, FB; Kalman, DA; Dills, R; Norman, D; Cain, ESK; Giabbai, M; Ghosal, M. Presence of Phenolic Compounds in Sewage Effluent and

Sludge from Municipal Sewage Treatment Plants. *Water Science and Technology* 1982, 14, 143-150.

[65] Rogers, KR; Gerlach, CL. Environmental Biosensors, A Status Report, *Environmental Science and Technology* 1996, 30 (11), 486A–491A.

[66] Paasivirta, J; Heinola, K; Humppi, T; Karjalainen, A; Knuutinen, J; Mäntykoski, K; Paukku, R; Piilola, T; Surma-Aho, K; Tarhanen, J; Welling, L; Vihonen, H; Särkkä, Polychlorinated Phenols, Guaiacols and Catechols in Environment. *Chemosphere* 1985, 14 (5), 469-491.

[67] ***EEC Directive 80-778-EEC 15-7-1990: Official Journal of the European Communities, 30-8-1990. European Community, Brussels.

[68] ***U.S. Environmental Protection Agency. Toxic Substance Control Act 1976.

[69] *** 2000/60/EC Directive of the European Parliament and of the Concil of 23 October 2000 establishing a framework of Community action in the field of water policy.

[70] Brega, A; Prandini, P; Amaglio, C; Pafuni, E. Determination of Phenol, *m*-, *o*- and *p*-cresol, *p*-aminophenol and *p*-nitrophenol in Urine by High-Performance Liquid Chromatography. *Journal of Chromatography A* 1990, 535, 311-316.

[71] Wissiak, R; Rosenberg, E. Universal Screening Method for the Determination of US Environmental Protection Agency Phenols at the Lower ng l^{-1} Level in Water Samples by On-line Solid-phase Extraction–high-performance Liquid Chromatography–atmospheric Pressure Chemical Ionization Mass Spectrometry within a Single Run. *Journal of Chromatography A* 2002, 963, 149-157.

[72] Kontsas, H; Rosenberg, C; Jappinen, P; Riekkola, MI. Glass Fibre Prefilter-XAD-2 Sampling and Gas Chromatographic Determination of Airborne Chlorophenols. *Journal of Chromatography A* 1993, 636(2), 255-261.

[73] Kontsas, H; Rosenberg, C; Pfaeffli, P; Jappinen, P. Gas chromatographic – Mass Spectrometric Determination of Chlorophenols in the Urine of Sawmill Workers with Past Use of Chlorophenol – Containing Anti – Stain Agents. *Analyst* 1995, 120(6), 1745 -1749.

[74] Tonokura, K; Nakamura, T; Koshi, M. Detection of Chlorobenzene Derivatives Using Vacuum Ultraviolet IonizationTime-of-Flight Mass Spectrometry. *Analytical Sciences* 2003, 19, 1109-1116.

[75] Ežerskis, Z; Jusys, Z. Oxidation of Chlorophenols on Pt Electrode in Alkaline Solution Studied by Cyclic Voltammetry, Galvanostatic

Electrolysis, and Gas Chromatography-Mass Spectrometry. *Pure Applied Chemistry* 2001, 73(12), 1929-1940.

[76] Gherardini, F; Michaud, PA; Panizza, M; Comninellis, Ch; Vatistas, N. Electrochemical Oxidation of 4-Chlorophenol for Wastewater Treatment. Definition of Normalized Current Efficiency (Φ). *Journal of the Electrochemical Society* 2001, 148 (6), D72-D82.

[77] Mailley, P; Cummings, EA; Mailley, S; Cosnier, S; Eggins, BR; McAdams, E. Amperometric Detection of Phenolic Compounds by Polypyrrole-Based Composite Carbon Paste Electrodes. *Bioelectrochemistry* 2004, 63, 291-296.

[78] Rodrigo, MA; Michaud, PA; Duo, I; Panizza, M; Cerisola, G; Comninellis, Ch.. Oxidation of 4-Chlorophenol at Boron-Doped Diamond Electrode for Wastewater Treatment. *Journal of the Electrochemical Society* 2001, 148 (5), D60-D64.

[79] Skowronski, JM; Krawczyk, P. Improved Electrooxidation of Phenol at Exfoliated Graphite Electrodes. *Journal of Solid State Electrochemistry* 2007, 11, 223-230.

[80] Sun, D; Zhang, H. Electrochemical Determination of 2-Chlorophenol Using an Acetylene Black Film Modified Carbon Electrode. *Water Research* 2006, 40, 3069-3074.

[81] Terashima, C; Rao, TN; Sarada, BV; Tryk, DA; Fujishima, A. Electrochemical Oxidation of Chlorophenols at a Boron-Doped Diamond Electrode and Their Determination by High-performance Liquid Chromoatography with Amperometric Detection. *Analytical Chemistry* 2002, 74, 895-902.

[82] Ureta-Zañartu, MS; Bustos, P; Berrios, C; Diez, MC; Mora, ML; Gutierrez, C. Electrooxidation of 2,4-dichlorophenol and Other Polychlorinated Phenols at a Glassy Carbon Electrode. *Electrochimica Acta* 2002, 47, 2399-2406.

[83] Polat, K. A Novel, Semi-pilot Scale Electrolysis System for the Production of p-Aminophenol. *Turkish Journal of Chemistry* 2003, 27, 501-511.

[84] DeLorenzo, ME; Scott, GI; Ross, PE. Toxicity of Pesticides to Aquatic Microorganisms: A Review. *Environmental Toxicology and Chemistry* 2000, 20, 84-98.

[85] Zhang, X; Wang, S; Shen, Q. The Electrochemical Behavior of p-Aminophenol at a ω-Mercaptopropionic Acid Self-Assembled Gold Electrode. *Microchimica Acta* 2005, 149, 37-42.

[86] Van Bocxlaer, JF; Clauwaert, KM; Lambert, WE; De Leenheer, AP. Quantitative Colorimetric Determination of Urinary p-Aminophenol With an Automated Analyzer. *Clinical Chemistry* 1997, 43, 627-634.
[87] Cañizares, P; Sáez, C; Lobato, J; Rodrigo, MA. Electrochemical treatment of 2,4-Dinitrophenol Aqueous Wastes using Boron-Doped Diamond Anodes. *Electrochimica Acta* 2004, 49 (26), 4641-4650.
[88] Karuwan, C; Mantim, T; Chaisuwan, P; Wilairat, P; Grudpan, K; Jittangprasert, P; Einaga, Y; Chailapakul, O; Suntornsuk, L; Anurukvorakun, O; Nacapricha, D. Pulsed Amperometry for Anti-fouling of Boron-Doped Diamond in Electroanalysis of β-Agonists: Application to Flow Injection for Pharmaceutical Analysis. *Sensors* 2006, 6, 1837-1850.
[89] Bebeselea, A; Proca, C; Manea, F; Radovan, C; Burtica, G; Shoonman, J. Electrochemical Degradation and Determination of 2,4-Dinitrophenol from Water. *Environmental Engineering and Management Journal* 2009, 8(4), 817-823.
[90] ***Health Council of the Netherlands: Committee on Updating of Occupational Exposure Limits. Oxalic Acid; Health-based Reassessment of Administrative Occupational Exposure Limits. The Hague: Health Council of the Netherlands, 2004; 2000/15OSH/106.
[91] Menache, R. Routine Micromethod for Determination of Oxalic Acid in Urine by Atomic Absorption Spectrophotometry. *Clinical Chemistry* 1974, 20, 1444-1445.
[92] Powers, H H; Levatin, P. A Method for the Determination of Oxalic Acid in Urine. *Journal of Biological Chemistry* 1944, 154, 207-213.
[93] Quisepe, C; Villasenor, J; Pecchi, G; Reyes, P. Catalytic Ozonation of Oxalic Acid with MnO_2/TiO_2 and Rh/TiO_2. *Journal of Chilean Chemical Soc*iety 2006, 51, 1049-1051.
[94] Casella, IG. Electrocatalytic Oxidation of Oxalic Acid on Palladium-Based Modified Glassy Carbon Electrode in Acidic Medium. *Electrochimica Acta* 1999, 44, 3353-3360.
[95] Martinez-Huitle, CA; Ferro, S; De Batisti, A. Electrochemical Incineration of Oxalic Acid. Role of Electrode Material. *Electrochimica Acta* 2004, 49, 4027-4034.
[96] Zarembski, PM; Hodgkinson, A. The Fluorimetric Determination of Oxalic Acid in Blood Biological Materials. *Biochemical Journal* 1965, 96, 717-721.
[97] Ensafi, AA; Emadi, M. Spectrophotometric Reaction Rate Method for Determination of Oxalic Acid in Food Based on Its Enhancing Effect on

the Oxidation of Pyrocathecol Violet by Dichromate. *Analytical Letter* 2004, 37, 321-332.

[98] Wilson, ChW; Shaw, PhE; Knight, RJ. Analysis of Oxalic Acid in Carambola (*Averrhoa carambola* L.) and Spinach by High-Performance Liquid Chromatography. *Journal of Agricultural and Food Chemistry* 1982, 30, 1106-1108.

[99] Gancedo, MC; Luh, BS. HPLC Analysis of Organic Acids and Sugars in Tomato Juice. *Journal of Food Sciences* 1986, 51, 571-573.

[100] Oderiz, V; Blanco, V; Hernandez, L; Lozano, S; Rodriguez, R. Simultaneous Determination of Organic Acids and Vitamin C in Green Beans by Liquid Chromatography. *Journal of AOAC International* 1994, 77, 1056-1059.

[101] Khaskhali, MH; Bhanger, MI; Khand, FD. Simultaneous Determination of Oxalic Acid and Citric Acids in Urine by High-Performance Liquid Chromatography. *Journal of Chromatography B. Biomedical Application* 1996, 675, 147-151.

[102] Zhanguo, C; Jiuru, L. Simultaneous and Direct Determination of Oxalic Acid, Tartaric Acid, Malic Acid, Vitamin C, Citric Acid, and Succinic Acid in Fructus Mume by Reversed-Phase High-Performance Liquid Chromatography. *Journal of Chromatography Science* 2002, 40, 35-39.

[103] He, Z; Yuan, L; Luo, Q; Zeng, Y. Simultaneous Determination of Oxalic and Tartaric Acid with Chemiluminescence Detection. *Analyst* 1997, 122, 1343-1345.

[104] Wu, F; He, Z; Luo, Q; Zeng, Y. High-Performance Liquid Chromatographic Determination of Oxalic Acid in Tea Using Tris (1,10-phenanthroline)-ruthenium (II) Chemiluminescence. *Analytical Science* 1998, 14, 971-973.

[105] Hong, F; Nilvebrant, N-O; Jonsson, LJ. Rapid and Convenient Determination of Oxalic Acid Employing a Novel Oxalate Biosensor Based on Oxalate Oxidase and SIRE Technology. *Biosensors and Biolectronics* 2003, 18, 1173-1181.

[106] Berna, A; Rodes, A; Feliu, JM. Oxalic Acid Adsorption and Oxidation at Platinum Single Crystal Electrodes. *Journal of Electroanalytical Chemistry* 2004, 563, 49-62.

[107] Pron'kin, SN; Petrii, OA; Tsirlina, GA; Schiffrin, DJ. Size Effects on the Electrochemical Oxidation of Oxalic Acid on Nanocrystalline Platinum. *Journal of the Electroanalytical Chemistry* 2000, 480, 112-119.

[108] McMurray, TA; Byrne, JA; Dunlop, PSM; Winkelman, JGM; Eggins, BR; McAdams, ET. Intrinsec Kinetics of Phoytocatalytic Oxidation of

Formic and Oxalic Acid on Immobilised TiO_2 Films. *Applied Catalysis A: General* 2004, 262, 105-110.

[109] Pera-Titus, M; Baños, G-M; Giménez, MA; Esplugas, S. Degradation of Chlorophenols by Means of Advanced Oxidation Processes: A General Review. *Applied Catalysis B: Environmental* 2004, 47, 219-256.

[110] Cañizares, P; García-Gómez, J; Sáez, C; Rodrigo, MA. Electrochemical Oxidation of Several Chlorophenols on Diamond Electrodes: Part II. Influence of Waste Charactersitics and Operating Conditions. *Journal of Applied Electrochemistry* 2004, 34, 87-94.

[111] Cañizares, P; Lobato, J; Rodrigo, MA; Sáez, C. Electrochemical Oxidation of Phenolic Waste With Boron-Doped Diamond Anodes. *Water Research* 2005, 39, 2687-2703.

[112] Pozniac, T; Araiza, B. Ozonation of Non-biodegradable Mixtures of Phenol and Naphthalene Derivates in Tanning Wastewaters. *Ozone: Science and Engineering* 2005, 27(5), 351-357.

[113] Boudenne, J-L; Cerclier, O. Performance of Carbon Black-slurry Electrodes for 4-Chlorophenol Oxidation. *Water Research* 1999, 33(2), 494-504.

[114] Luz, RCS; Damos, FS; de Oliveira, AB; Beck, J; Tatsuo, L. Voltammetric Determination of 4-Nitrophenol at a Lithium Tetracyanoethylenide (LiTCNE) Modified Glassy Carbon Electrode. *Talanta* 2004, 64, 935-942.

[115] Pedrosa, VA; Codognoto, L; Avaca, LA. Electroanalytical Determination of 4-Nitrophenol by Square Wave Voltammetry on Diamond Electrodes. *Journal of Brazilian Chemical Society* 2003, 14, 530-538.

[116] Charoenraks, T; Chuanuwatanakul, S; Honda, K; Yamaguchi, Y; Chailapakul, O. Analysis of Tetracycline Antibiotics using HPLC with Pulsed Amperometric Detection. *Analytical Sciences* 2005, 21, 241-245.

[117] Carvalho, RH; Lemos, F; Lemos, MANDA; Cabral, MS; Ramôa Ribeiro, F. Electro-oxidation of Phenol on Zeolite/graphite Composite Electrodes: Part I. Electrochemical Behaviour Over NaY Zeolite. *Journal of Molecular Catalysis A: Chemical* 2006, 248, 48-52.

[118] Carvalho, RH; Lemos, MANDA; Lemos, F; Cabral, JMS; Ramôa Ribeiro, F. Electro-oxidation. *Journal of Molecular Catalysis A: Chemical* 2006, 253, 170-176.

[119] Carvalho, RH; Lemos, F; Lemos, MANDA; Cabral, JMS; Ramôa Ribeiro, F. Electro-oxidation of Phenol on a New Type of

Zeolite/Graphite Biocomposite Electrode with Horseadish Peroxidase. *Journal of Molecular Catalysis A: Chemical* 2007, 278, 47.

[120] Walcarius, A; Rozanska, S; Bessiére, J; Wang J. Screen-Printed Zeolite-Modified Carbon Electrodes. *Analyst* 1999, 124, 1185-1190.

[121] Behrendt, J; Arevalo, E; Gulyas, H; Nlederste-Hollenberg, J; Niemiec, A; Zhou, J; Otterpohl, R. Production of Value Added Products from Separately Collected Urine. *Water Science and Technology* 2002, 46, 341-346.

[122] Iida, Y; Suganuma, Y; Matsumoto, K; Satoh, I. Novel Determination System for Urea in Alcoholic Beverages by Using an FIA System With an Acid Urease Column. *Analytical Science* 2006, 22, 173-176.

[123] Ferrer, SJ; Granados, SG; Bedioui, F; Ordaz, AA. Amperometric Detection of Urea in Aqueous Solution by Poly(Ni-Cyclam) Film-Modified Glassy Carbon Electrode. *Electroanalysis* 2003, 15, 70-73.

[124] Jin, J-H; Paek, S-H; Lee, C-W; Min, NK; Hong, S-I. Fabrication of Amperometric Sensor Based on Nano-Porous Silicon Technology. *Journal of Korean Physical Society* 2003, 42, 735-738.

[125] Yang, J-K; Ha, KS; Baek, HS; Lee, SS; Seo, ML. Amperometric Determination of Urea Using Enzyme-Modified Carbon Paste Electrode. *Bulletin of Korean Chemical Society* 2004, 25, 1499-1502.

[126] Veeneman, JM; Kingma, HA; Stellard, F; de Jong, PE; Reijngoud, D-J; Huisman, R.M. Comparison of Amino Acid Oxidation and Urea Metabolism in Haemodialysis Patients During Fasting and Meal Intake. *Nephrology Dialysis Transplantation* 2004, 19, 1533-1541.

[127] Eggenstein, C; Borchardt, M; Diekmann, C; Gründig, B; Dumscaht, C; Cammann, K; Knoll, M; Spener, F. A Disposable Biosensor for Urea Determination in Blood Based on a Ammonium-Sensitive Transducer. *Biosensors and Biolectronics* 1999,14, 33-41.

[128] Walcerz, I; Stanislaw, G; Koncki, R. Potentiometric Enzyme Electrode in a Flow Injection System for the Determination of Urea in Human Serum Samples. *Analytica Chimica Acta* 1998, 369, 129-137.

[129] Hampson, NA; Lee, JB; Morley, JR; Scanlon, B. Oxidations Including Silver.V. The Oxidation of Primary Aliphatic Amines. *Canadian Journal of Chemistry* 1969, 47, 3729-3736.

[130] Kamin, RA; Wilson, GS. Rotating Ring-disk Enzyme Electrode for Biocatalysis Kinetic Studies and Characterization of the Immobilized Enzyme Layer. *Analytical Chemistry* 1980, 52, 1198-1205.

[131] Stulik, K; Pacakova, V. Amperometric Flow Detection with a Copper Working Electrode-Response Mechanism and Application to Various Compounds. *Talanta* 1988, 35(6), 455-460.

[132] Torto, N; Ruzgas, T; Gorton, L. Electrochemical Oxidation of Mono- and Disaccharides at Fresh as well as Oxidized Copper Electrodes in Alkaline Media, *Journal of Electroanalytical Chemistry* 1999, 464, 252-258.

[133] Marioli, JM; Kuwana, T. Electrochemical Characterization of Carbohydrate Oxidation at Copper Electrodes. *Electrochimica Acta* 1992, 37(7), 1187-1197.

[134] Casella, IG; Gatta, M. Determination of Electroactive Organic Acids by Anion-exchange Chromatography Using a Copper Modified Electrode. *Journal of Chromatography A* 2001, 912, 223-233.

[135] Chen, Z-L; Hibbert, DB. Simultaneous Amperometric and Potentiometric Detection of Sugars, Polyols and Carboxylic Acids in Flow Systems Using Copper Wire Electrodes. *Journal of Chromatography A* 1997, 766, 27-33.

[136] Luo, MZ; Baldwin, RP. Characterization of Carbohydrate Oxidation at Copper Electrodes. *Journal of Electroanalytical Chemistry* 1995, 387, 87-94.

[137] Mho, S; Johnson, DC. Electrocatalytic Response of Carbohydrates at Copper-Alloy Electrodes. *Journal of Electroanalytical Chemistry* 2001, 500, 524-532.

[138] Nagy, L; Nagy, G; Hajos, P. Copper Electrode Based Amperometric Detector Cell for Sugar and Organic Acid Measurements. *Sensors and Actuators B* 2001, 76, 494-499.

[139] Prabhu, SV; Baldwin, RP. Constant Potential Amperometric Detection of Carbohydrates at a Copper-Based Chemically Modified Electrode, *Analytical Chemistry* 1989, 61, 852-856.

[140] Cataldi, TRI; Centonze D. Development of a Carbon Composite Electrode Made from Polyethylene and Graphite Powder Modified with Copper(I) Oxide. *Analytica Chimica Acta*, 1996, *326*, 107-115.

INDEX

D

E

F

G

H

I

K

T

U

V

W

Z